**W9-CHF-350**

# Low-Dimensional Structures in Semiconductors
## From Basic Physics to Applications

# NATO ASI Series

## Advanced Science Institutes Series

*A series presenting the results of activities sponsored by the NATO Science Committee,*
*which aims at the dissemination of advanced scientific and technological knowledge,*
*with a view to strengthening links between scientific communities.*

The series is published by an international board of publishers in conjunction with the
NATO Scientific Affairs Division

| | | |
|---|---|---|
| A | **Life Sciences** | Plenum Publishing Corporation |
| B | **Physics** | New York and London |
| | | |
| C | **Mathematical and Physical Sciences** | Kluwer Academic Publishers |
| D | **Behavioral and Social Sciences** | Dordrecht, Boston, and London |
| E | **Applied Sciences** | |
| | | |
| F | **Computer and Systems Sciences** | Springer-Verlag |
| G | **Ecological Sciences** | Berlin, Heidelberg, New York, London, |
| H | **Cell Biology** | Paris, Tokyo, Hong Kong, and Barcelona |
| I | **Global Environmental Change** | |

*Recent Volumes in this Series*

*Series B: Physics*

# Low-Dimensional Structures in Semiconductors
## From Basic Physics to Applications

Edited by

# A. R. Peaker

University of Manchester Institute of Science and Technology
Manchester, United Kingdom

and

# H. G. Grimmeiss

University of Lund
Lund, Sweden

Plenum Press
New York and London
Published in cooperation with NATO Scientific Affairs Division

Proceedings of a NATO Advanced Study Institute/Nineteenth Course of the
International School of Materials Science and Technology
on Low-Dimensional Structures in Semiconductors:
From Basic Physics to Applications,
held July 1–14, 1990,
in Erice, Sicily, Italy

Library of Congress Cataloging-in-Publication Data

Low-dimensional structures in semiconductors : from basic physics to
applications / edited by A.R. Peaker and H.G. Grimmeiss.
      p.   cm. -- (NATO ASI series. Series B, Physics ; v. 281)
   Reviews presented at the NATO Advance Study Institute on "Low
Dimensional Structures in Semiconductors: from Basic Physics to
Applications," held in Erice, Sicily, Italy, Jul. 1-14, 1990.
   "Published in cooperation with NATO Scientific Affairs Division."
   Includes bibliographical reference and indexes.
   ISBN 0-306-44086-5
   1. Semiconductors--Surfaces--Congresses.  2. Layer structure
(Solids)--Congresses.   I. Peaker, A. R.  II. Grimmeiss, H. G.
III. NATO Advanced Study Institute on "Low Dimensional Structures in
Semiconductors: from Basic Physics to Applications" (1990 : Erice,
Italy)  IV. Series.
QC611.6.S9L68  1992
537.6'221--dc20                                        91-40001
                                                          CIP

ISBN 0-306-44086-5

© 1991 Plenum Press, New York
A Division of Plenum Publishing Corporation
233 Spring Street, New York, N.Y. 10013

Printed in the United States of America

# PREFACE

This volume contains a sequence of reviews presented at the NATO Advanced Study Institute on 'Low Dimensional Structures in Semiconductors ... from Basic Physics to Applications.' This was part of the International School of Materials Science and Technology held in July 1990 at the Ettore Majorana Centre in Sicily.

Only a few years ago, Low Dimensional Structures was an esoteric concept, but now it is apparent they are likely to play a major role in the next generation of electronic devices. The theme of the School acknowledged this rapidly developing maturity. The contributions to the volume consider not only the essential physics, but take a wider view of the topic, starting from material growth and processing, then progressing right through to applications with some discussion of the likely use of low dimensional devices in systems.

The papers are arranged into four sections, the first of which deals with basic concepts of semiconductor and low dimensional systems. The second section is on growth and fabrication, reviewing MBE and MOVPE methods and discussing the achievements and limitations of techniques to reduce structures into the realms of one and zero dimensions. The third section covers the crucial issue of interfaces while the final section deals with devices and device physics.

The meeting attendance approached a hundred, but many more students and research scientists applied to come, sadly we could not accommodate them all. We hope that this book will be of value not only to those who attended but also to a wider audience in providing a permanent record of some of the lectures given during the two weeks of the School.

We would like to thank the many people who put so much effort into making the meeting a success. In particular, our secretaries at UMIST and Lund, Jacqui Gilmore and Mona Hammar, the staff at Erice, J. Pilarski, Pinola Savalli, Caterina Greco and their assistants. We would also like to thank NATO for funding this ASI and for additional support from Dr. A. Zichichi, Director of the Erice Centre, the Italian Ministries of Education and Science Research, the Sicilian Regional Government, the Swedish National Research Council, the Swedish National Board for Technical Development, the European Physical Society, ABB Hafo, AB LM Ericsson, the University of Lund and the University of Manchester Institute of Science and Technology.

*The Organising Committee*

*Professor M Balkanski, University Pierre & Marie Curie, Paris, France*
*Dr A Gabrielle, Ettore Majorana Centre, Erice, Sicily, Italy*
*Professor H G Grimmeiss, University of Lund, Sweden*
*Professor A R Peaker, UMIST, Manchester, UK*

# CONTENTS

## CONCEPTS

## GROWTH AND FABRICATION

## INTERFACES

## DEVICES

## INDEXES

# THE VALENCE SUB-BANDS OF BIASED SEMICONDUCTOR HETEROSTRUCTURES

R. Ferreira and G. Bastard

Laboratoire de Physique de la Matière Condensée
l'Ecole Normale Supérieure, 24 rue Lhomond
F-75005 Paris, France

It is now recognised that the envelope function framework[1-3] is a versatile and convenient method to calculate the electronic states of semiconductor heterostructures. In essence this computational scheme relies on the separation between the rapidly varying periodic parts of the Bloch functions, assumed to be the same in each kind of layer, and the more slowly varying envelope functions. The latter experience the potentials arising from the band bending due to charges and the band edge steps when one goes from one layer to the next. Thus, a great deal of the computational complexities is of bulk like origin, i.e. depends upon the fact that the extrema around which one is interested to build the heterolayer eigenstates is non degenerate or degenerate, the local effective mass tensor is scalar or not etc. In this respect, for III-V or II-VI heterolayers which retain the zinc blende lattice, there is a marked difference between the conduction and valence bands. The former is non degenerate (apart from spin) and displays small anisotropy and non parabolicity. The valence band edge instead is fourfold degenerate at the zone center ($\Gamma_8$ symmetry), which leads to complicated dispersion relations (although quadratic upon the wavevector $\mathbf{k}$)[4].

As a result of the simplicity of the host's conduction band edge one may readily calculate the conduction sub-bands by solving for each $k_\perp$ a scalar Schrôdinger equation:

$$\left[ p_z [1/2m(z)]p_z + \hbar^2 k_\perp^2/2m(z) + V_S(z) + V_{ext}(z) \right] \chi(z) = \varepsilon \chi(z) \quad (1)$$

where $V_S(z)$ is the position dependent conduction band edge ( 0 in the well acting material, $V_S$ in the barrier acting material), $m(z)$ the position dependent conduction band effective mass ($m_w$ in the well, $m_b$ in the barrier), $k_\perp = (k_x, k_y)$ and $V_{ext}(z)$ an external potential (arising e.g. from charges or an external electric field). When $k_\perp$ is small to allow the second term of eq.(1) to be treated perturbatively one finds the eigenenergies organise into two dimensional sub-bands:

$$\varepsilon_n(k_\perp) = E_n + \hbar^2 k_\perp^2 / 2m_n \tag{2}$$

where $m_n$ is such that

$$1/m_n = 1/m_w P_w + 1/m_b P_b \tag{3}$$

and $P_w$, $P_b$ are the integrated probabilities of finding the electron in the well and in the barrier respectively while in the $n^{th}$ state (confinement energy $E_n$). The complete wavefunction associated with eq.(2) is:

$$\psi_n(\mathbf{r}) = u_{c0}(\mathbf{r}) \chi_n(z) \exp(ik_\perp . r_\perp) / \sqrt{S} \tag{4}$$

where $r_\perp$ denotes the electron in-plane position, $u_{c0}(\mathbf{r})$ is the periodic part of the Bloch function at the zone center for the hosts' $\Gamma_6$ edge.

## VALENCE SUB-BANDS AT ZERO ELECTRIC FIELD

The bulk topmost valence hamiltonian is a 4x4 matrix, first derived by Luttinger[4], whose basis vectors can be identified with the four components of a J=3/2 angular momentum. In the envelope function approximation, the envelope eigenfunctions (which are 4x1 spinors on the |3/2, $m_J$ > basis where $-3/2 \leq m_J \leq 3/2$) of the Luttinger hamiltonian of a multiple quantum well structure are the solutions of:

$$\sum_{l'} H_{ll'} \psi_{l'}(z) = 0 \tag{5}$$

where $H_{ll'}$ is a 4x4 matrix whose elements , as written on the basis |3/2,3/2>; |3/2,-1/2>; |3/2,+1/2>; |3/2,-3/2>, are:

$$
\begin{pmatrix}
H_{hh} + V_P(z) & c & b & 0 \\
c^* & H_{lh} + V_P(z) & 0 & -b \\
b^* & 0 & H_{lh} + V_P(z) & c \\
0 & -b^* & c^* & H_{hh} + V_P(z)
\end{pmatrix}
\quad (6)
$$

where:

$$k_\perp = (k_\perp \cos\theta, k_\perp \sin\theta) \tag{7}$$

$$H_{hh} = -p_z(\gamma_1 - 2\gamma_2)p_z/2m_0 - \hbar^2 k_\perp^2(\gamma_1 + \gamma_2)/2m_0 \tag{8}$$

$$H_{lh} = -p_z(\gamma_1 + 2\gamma_2)p_z/2m_0 - \hbar^2 k_\perp^2(\gamma_1 - \gamma_2)/2m_0 \tag{9}$$

$$c = \hbar^2\sqrt{3}\gamma_2 k_\perp^2\, e^{-2i\theta}/2m_0 \tag{10}$$

$$b = -i\sqrt{3}\hbar^2 k_\perp\, e^{-i\theta}/2m_0\, [\gamma_3 d/dz + d/dz\gamma_3] \tag{11}$$

while $V_P(z)$ represents the position dependent valence band edge and $\gamma_1, \gamma_2, \gamma_3$ are the Luttinger parameters of the bulk materials. In eq.(10) the axial approximation[1] has been used. It renders the valence dispersions isotropic in the layer plane. It is interesting to note that the off diagonal elements of eq.(6) vanish at $k_\perp = 0$. Thus, a convenient method to find the $k_\perp$ dependence of the eigenstates consists of diagonalising these off diagonal elements in a basis spanned by the $k_\perp = 0$ solutions of eq.(6) . This basis includes the bound states and the continuum states of the problem. Since one is usually interested in knowing only the topmost dispersion relations, it is a sensible approximation to discard the continuum states and to retain only the $k_\perp = 0$ bound states. This incomplete basis has however the drawback of not complying with the conservation of the probability current at the interfaces. The latter can be expressed by integrating eq.(6) once with respect to z across an interface. Since the off diagonal terms of eq.(6) involve the $\gamma_2$, $\gamma_3$ parameters , which are in principle position dependent, the conservation of the probability current involves a 4x4 matrix $C_{ll'}$ which is not diagonal. The use of a finite $k_\perp = 0$ basis does nor allow to cope with the off diagonal terms of $C_{ll'}$. In practice however, the penetration of the topmost valence levels outside the well is faint and the violation of the current conserving conditions is of minor importance. One noticeable exception to this rule of thumb is found in the HgTe-CdTe system where the differences in the $\gamma_1, \gamma_2$ parameters are large ( $\gamma_1$ and $\gamma_2$ are of opposite signs in the two materials) and the penetration in the CdTe layers is non negligible due to the light conduction band mass in HgTe (0.03 $m_0$).

At $k_\perp = 0$ the eigenstates of eq.(6) split into two decoupled sets of levels. Those corresponding to $m_J = \pm 3/2$ (respectively $\pm 1/2$) correspond to heavy (light) holes since the respective confinement energies involve a heavy $[\, m_0/(\gamma_1 - 2\gamma_2)]$ and a light $[\, m_0/(\gamma_1 + 2\gamma_2)]$ effective mass. We shall denote these $k_\perp = 0$ edges by $HH_n$ (respectively $LH_n$) where $n \geq 1$. If the b and c terms were always discarded the in plane dispersions of the valence levels would exhibit the "mass reversal effect". Namely, the in plane effective masses would be lighter for the $HH_n$ sub-bands $[\, m_0/(\gamma_1 + \gamma_2)]$ than for the $LH_n$ ones $[\, m_0/(\gamma_1 - \gamma_2)]$. This would lead to the crossings of the heavy and light hole branches, for instance at $k_1 = [\, m_0( HH_1 - LH_1 )/ \gamma_2 \hbar^2]^{1/2}$ for the topmost sub-bands. The non vanishing off diagonal terms replace these crossings by anticrossings. The latter feature implies strongly non parabolic in plane dispersion relations . This is particularly noticeable for the $LH_1$ sub-bands which acquire a camel back shape near $k_\perp = 0$. Large mixing of the heavy and light hole characters have to be expected when the anticrossings take place and in fact the very notion of heavy and light hole becomes fuzzy at non vanishing $k_\perp$. Such considerations can be quantified by evaluating the $k_\perp$ dependence of the average value of $J_z$ over the various eigenstates. As will be shown below these averages deviate quickly from the edge values ($\pm 3/2$ for $HH_n$, $\pm 1/2$ for $LH_n$).

It is actually possible to derive the in plane dispersion relations and the corresponding eigenfunctions of a single quantum well under flat band conditions (F = 0) almost analytically if one restricts the $k_\perp =$ basis to the $HH_1$, $HH_2$ and $LH_1$ edges[5-7]. Under these assumptions the dispersion relations are the roots of:

$$(\varepsilon_{HH1} - \varepsilon)[( \varepsilon_{LH1} - \varepsilon)( \varepsilon_{HH2} - \varepsilon) - |\langle\phi_1| b | \chi_2\rangle|^2 ] - |\langle\phi_1| c | \chi_1\rangle|^2 (\varepsilon_{HH2} - \varepsilon) = 0 \qquad (12)$$

where:

$$\varepsilon_{HHi} = HH_i - \hbar^2 k_\perp^2( \gamma_1 + \gamma_2)/2m_0 \qquad (13)$$
$$\varepsilon_{LHi} = LH_i - \hbar^2 k_\perp^2( \gamma_1 - \gamma_2)/2m_0 \qquad (14)$$

and $\phi_1$, $\chi_1$, $\chi_2$ are the quantum well envelope functions of the $LH_1$, $HH_1$ and $HH_2$ states at $k_\perp = 0$. Each of these eigenvalues are twice degenerate (Kramers degeneracy) and to each of the twofold degenerate eigenenergies one can asociate the two orthogonal wavefunctions:

$$\psi_{k_\perp\uparrow} = 1/\sqrt{S} \ [1+a^2+\eta^2]^{(-1/2)} \ e^{ik_\perp \cdot r_\perp}$$
$$x[a \ e^{-2i\theta} \chi_1(z), \ \phi_1(z), \ 0, \ i\eta \ e^{i\theta} \chi_2(z)] \tag{15}$$
$$\psi_{k_\perp\downarrow} = 1/\sqrt{S} \ [1+a^2+\eta^2]^{(-1/2)} e^{ik_\perp \cdot r_\perp}$$
$$x \ [- \ i\eta \ e^{-i\theta} \chi_2(z), \ 0, \ \phi_1(z), \ a \ e^{2i\theta} \chi_1(z) \ ] \tag{16}$$

where:

$$a = \hbar^2 \sqrt{3} k_\perp^2 \ \gamma_2 < \phi_1 | \chi_1 > /[2m_0(\varepsilon - \varepsilon_{HH1})] \tag{17}$$

$$\eta = \hbar^2 \sqrt{3} k_\perp < \chi_2 \ | \ \gamma_3 d/dz + d/dz\gamma_3 | \ \phi_1 > /[2m_0(\varepsilon - \varepsilon_{HH2})] \tag{18}$$

One readily checks that $< \psi_{k_\perp\uparrow} | J_z | \psi_{k_\perp\uparrow}>$ (respectively $< \psi_{k_\perp\downarrow} | J_z |$ $\psi_{k_\perp\downarrow}>$) extrapolates to +3/2 (respectively -3/2) for the lowest lying hole states (the HH$_1$ branch). Moreover, $J_z$ has no non-vanishing matrix element between any $\psi_{k_\perp\uparrow}$ and $\psi_{k_\perp\downarrow}$ corresponding to the same energy . This has led us to label the hole eigenstates according to the "spin" ($\downarrow,\uparrow$) eventhough neither $\psi_{k_\perp\uparrow}$ nor $\psi_{k_\perp\downarrow}$ are eigenstates of $\sigma_z$ or $J_z$ at finite $k_\perp$.

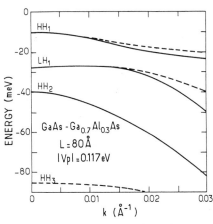

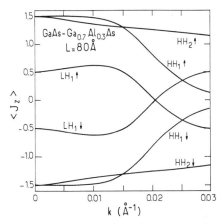

Fig(1a). Calculated in-plane dispersion relations of the topmost valence sub-bands in a 80 Å thick GaAs-Ga$_{0.7}$Al$_{0.3}$As single quantum well. Solid line: three level model. Dashed lines: inclusion of all the levels bound in the well at k = 0 .

Fig(1b). Calculated in-plane dependence of the J$_z$ averages over the energy eigenstates in the three level model. The notations $\uparrow$, $\downarrow$ refer here to the dominant character at k = 0. They agree with (are opposite to)those used in the text for the HH$_1$ (LH$_1$,HH$_2$) sub-bands.

Figures (1a,b) show the calculated dispersion relations and averages of J$_z$ over the $\uparrow$ and $\downarrow$ eigenstates for a 80Å GaAs-Ga(Al)As quantum well (valence band offset $|Vp| = 117$ meV, $\gamma_1 = 6.85$, $\gamma_2 = 2.1$, $\gamma_3 = 2.9$ respectively)[7]. As mentionned previously, the $J_z$ averages deviate from $\pm 3/2$ ($\pm 1/2$) for the heavy (light) hole branches when they anticross. We also show for comparison in fig.(1a) the dispersion curves obtained by retaining more sub-bands in the $k_\perp = 0$ basis. It is seen that over significant $k_\perp$ range the three level model gives a fair account of the HH$_1$ dispersion.

Knowing the dispersion relations and the wavefunctions one can calculate a number of physical properties such as for instance the level lifetimes associated with defects, phonons etc...A particular class of relaxation effects are those which correspond to spin flips of the carrier. They are very fast for holes in bulk materials, the reason being that the spin-orbit effects are very large: a hole created at a finite **k** value with a given "spin", as obtained by quantizing **J** along **k**, will almost immediatly lose its "spin" by any scattering mechanism, which because it implies a change in the hole wavevector also implies, in the final state, a different "spin" quantum number. Such a fast hole "spin" flip becomes inhibited in quantum wells, at least if $k_\perp$ is not too large.

The reason for this quenching is the lifting of the heavy and light hole degeneracy by the quantum well potential which makes the HH$_1$ states with a small $k_\perp$ to be nearly $\pm 3/2$ states, which cannot be coupled by any static potential. An example of such a behaviour is shown in fig.(2) where the "spin"-flip and "spin"-conserving scattering times of holes due to ionised impurity or alloy fluctuations is plotted versus $k_\perp$ for bulk $Ga_{0.47}In_{0.53}As$ or $Ga_{0.47}In_{0.53}As$ - InP quantum wells[7].

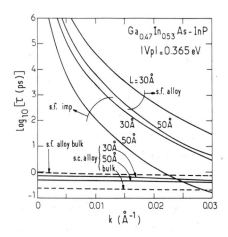

Fig.(2). Calculated dependence upon the in-plane wavevector of the impurity- and alloy fluctuations-assisted "spin"-flip (s.f.) and "spin"-conserving (s.c.) scattering times within the HH$_1$ sub-band of $Ga_{0.47}In_{0.53}As$-InP quantum wells and bulk $Ga_{0.47}In_{0.53}As$.

Two salient features are noticeable. Firstly, the "spin"-conserving scattering is much faster than the "spin"-flip one in quantum wells. Also, it is almost independent of the in-plane wavevector and of the quantum well thickness: the bulk and quantum well figures are of comparable magnitudes. In contrast, the "spin"-flip relaxation times are strongly $k_\perp$ dependent in quantum wells and over a significant $k_\perp$ range may be longer than the recombination time (500 ps - 1 ns). The "spin"-flip relaxation time of holes in bulk $Ga_{0.47}In_{0.53}As$ is instead very short and $k_\perp$ independent. Thus, there is a clear effect asssociated with the size quantization upon the strength of some scattering mechanisms. The spin relaxation times play a key part in the spin orientation experiments where a circularly polarized light creates a preferential spin orientation of the photo-generated electrons and holes. This magnetisation decays due to the spin-flip relaxation and eventually the emitted light displays a residual degree of circular polarisation. Recently, Uenoyama and Sham[8] have succeeded in interpreting spin orientation measurements performed in doped and undoped GaAs-Ga(Al)As quantum wells by assuming that , instead of an immediate "spin"-relaxation, the holes keep some memory of their initial polarisation. The previous discussions have explained why it should be so from considerations of the valence sub-band structure.

The band mixing effects at finite $k_\perp$ has some important consequences upon the "vertical" transport of holes in a superlattice. Namely, it enhances the hole capability of hopping from one well to the other when $k_\perp$, i.e. the in-plane velocity increases. The physics of this effect can be simply understood upon examination of eq.(6). At $k_\perp = 0$ the heavy hole sub-band width for the z motion depends upon the transfer integral from one well to the other. The latter behaves like exp- $[2m_{hh}/\hbar^2(|V_P| + \varepsilon_{HH1})L_B]$ where $L_B$ is the barrier thickness and $m_{hh}/m_0 = (\gamma_1 - 2\gamma_2)^{-1}$ is the heavy hole effective mass along the growth axis. The ground light hole sub-band width is given by a similar expression where $m_{lh}$ ( $= m_0/(\gamma_1 + 2\gamma_2)$ ) and $\varepsilon_{LH1}$ replace $m_{hh}$ and $\varepsilon_{HH1}$ respectively, i.e. the tunnelling length for the light hole is longer than that of the heavy hole. At finite $k_\perp$ the heavy and light hole characters become mixed. Hence, one should expect the heavy hole motion along the growth axis to be characterised by a lighter effective mass when $k_\perp$ increases. Therefore, the bandwidth for the ground superlattice sub-band should increase upon $k_\perp$. This trend is demonstrated in fig.(3) where we show the $k_\perp$ dependence of the ground valence sub-band width in

GaAs-Ga$_{0.7}$Al$_{0.3}$As superlattices with equal layer thicknesses and three different periods[9]. A twofold increase of the bandwidth is found upon the k$_\perp$ increase from zero (no band mixing) to $3\times10^6$cm$^{-1}$ and d= 60 Å.

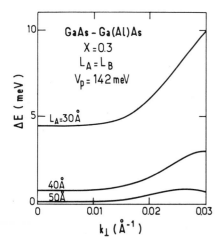

Fig.(3). Calculated k$_\perp$ dependence of the ground valence sub-band width in GaAs-Ga$_{0.7}$Al$_{0.3}$As superlattices with equal layer thicknesses and d = 60Å, 80Å and 100Å.

## ELECTRIC FIELD EFFECTS

The dominant effects on the valence eigenstates of single quantum wells which are associated with an electric field **F** applied parallel to the growth axis are[1-3,9] i) the spatial polarisation of the eigenstates and ii) the lifting of the Kramers degeneracy at finite k$_\perp$.

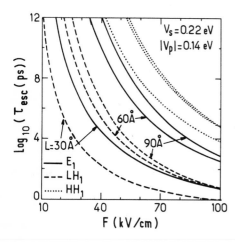

Fig.(4). Semiclassical estimates of the escape time of an electron (in the E$_1$ state), a heavy hole (in the HH$_1$ state) and a light hole (in the LH$_1$ state) at k$_\perp$= 0 out of quantum wells (L = 30Å, 60Å, 90Å) tilted by an electric field. For electron, light hole and heavy hole the escape time increases monotically with L.

The polarisation effects also show up in energy shifts ( due to the interaction of the induced dipole with the field) which are red shifts (quadratic upon F at low enough F) for the ground states HH$_1$ and LH$_1$ at $k_\perp = 0$. The lifting of the Kramers degeneracy occurs due to the combined effects of the non zero $k_\perp$, the finite spin-orbit energy and the non centro-symmetry of the tilted quantum well. Note that the field in principle prevents the existence of truly bound states. However, under most circumstances, the field induced ionisation can be safely neglected (see fig.(4) for a relevant example[9]) as the finite lifetime of the virtual bound states remains much longer than the period of the carrier oscillations inside the tilted quantum well.

In mutiple well structures the electric field can cause either a spatial localisation of states which were delocalised at F = 0 (the Wannier-Stark quantisation[10]) or induces a partial delocalisation when the potential energy drop over p periods peFd match the energy difference between two levels essentially localised in different wells ( field induced resonant tunnelling[11]). The latter mechanism is the one by which the holes coherently move along the growth axis of the structure while preserving their in plane momentum. This motion is actually negligible with respect to the one induced by the defects (or phonons): one may in fact show[12] that the coherent tunnelling current in a finite and biased multiple quantum well structure collapses with the field, the reason being that the field in general misaligns the consecutive hole levels and only exceptionnally aligns them. Once we abandon the coherent resonant tunnelling picture and allow for the existence of relaxation mechanisms, the misalignment of consecutive eigenstates can be circumvented by external scatterers. Firstly the longitudinal motion no longer requires the in plane wavevector to be conserved. This means that the scatterers convert some part of the initial longitudinal carrier energy into transverse kinetic energy in the final states while conserving the overall carrier energy if they are elastic or allowing an energy relaxation of the carrier if they are inelastic. Thus, we can define an assisted time from one hole eigenstate $|\alpha, k_\perp\rangle$ to all other hole eigenstates $|\beta, k'_\perp\rangle$ by using the Fermi golden rule:

$$\hbar/2\pi\tau_{\alpha k\perp} = \sum_{k'_\perp} |\langle \alpha k_\perp | V_{def} | \beta k'_\perp\rangle|^2 \, \delta(\varepsilon_{\alpha k\perp} - \varepsilon_{\beta k'\perp}) \quad (19)$$

It is clear from eq.(19) that the elastic assisted tunnelling should exhibit resonances at the same electric field values as the purely coherent tunnelling. In fact, when two eigenstates become aligned by the field their wavefunctions spatially delocalise. The assisted tunnelling rate is proportional to the square of the matrix element of the scattering potential between the initial and final states of the transition . Clearly when these are delocalised the strength of the coupling is greatly enhanced compared with the non resonant case, hence the resonance. This qualitative discussion can be sustained by calculations as will be shown below.

In the context of coupled double barriers, it is now widely recognised that one deals with assisted (also called sequential[13]) tunnelling from the emitter to the well and from the well to the collector. On the other hand , there have been attempts to interpret the experiments performed on coupled double wells in terms of the coherent (Rabi) oscillations that a carrier undergoes between the two wells when it is prepared in a quantum state which is not an eigenstate of the system. The observation of Rabi oscillations is easily achievable in the radio-frequency range where many quasi degenerate atomic states fall. On the other hand, in the context of laser excited double quantum wells, it seems to us hardly reasonable to imagine a smaller level splitting than the laser width (to obtain a coherent population of the two interacting states)   yet larger than the broadening of the eigenstates (to be able to talk about distinct eigenstates of the two well system). If we thus abandon the idea of Rabi oscillations, we have to admit that the carrier will only be created in eigenstates of the double quantum well, in which they will stay for ever if there is no scattering to enable them to make transitions. Thus, we believe that the appropriate framework of interpretation of the tunnelling in double quantum wells is at the present time the same as that prevailing in double barrier structures, i.e. that of the assisted tunnelling.

The coherent or assisted tunnellings of holes[14,15] nicely display the band mixing effects that were previously discussed: if heavy and light holes were always decoupled there would exist no possible conversion of a heavy to a light hole due to  static scatterers. Band mixing effects instead allow for such a coupling and thus for resonances in the assisted hole transfer when the electric field lines up heavy and light hole states. In general the band mixing effects accelerate the transfer over estimates based on decoupled heavy and light hole models. We illustrate this point in fig.(5) where we show the calculated impurity assisted hole transfer time versus the electric field strength in the case of a symmetrical GaAs-Ga(Al)As double quantum well[14]. The initial hole state was taken as $HH'_1$ ,the edge of the topmost valence level in the right hand side well. The solid line corresponds to the full calculations while the dashed line is based on a model which does not take the band mixing into account (b = c = 0 in the Luttinger matrix).

The level scheme at $k_\perp = 0$ is shown in fig.(6) and allows to identify the resonance at -70 kV/cm as due to the anticrossing between HH'$_1$ and HH$_2$ and that near -35 kV/cm to the "frustrated" anticrossing between HH'$_1$ and LH$_1$. The anticrossing is "frustrated" because it does not occur at $k_\perp = 0$, but only for non vanishing $k_\perp$. It is striking that over a large electric field range, calculations which neglect the band mixing overestimate the transfer time by many decades.

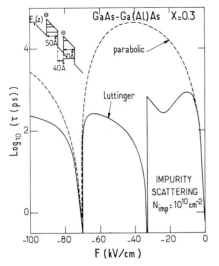

Fig.(5). The decimal logarithm of the impurity-assisted inter-well hole transfer time in a 50Å-40Å-50Å GaAs-Ga$_{0.7}$Al$_{0.3}$As biased double well is plotted versus the field strength F. The initial state is HH'$_1$.

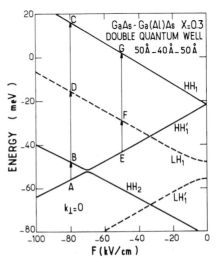

Fig.(6). Level scheme at $k_\perp = 0$ versus F for the heterostructure of fig.(5)

Compared to electronic resonances the hole resonances are narrower and deeper around the field strength which realises the anticrossing between the two interacting levels. This is due to the smaller hole energy splitting at resonance. The latter feature implies that, exactly at resonance, the in-plane wave vector that the scatterer has to provide to the tunnelling particle is smaller for a hole than for an electron. Since in general the scattering matrix element decreases with increasing in-plane wavevector transfer (see e.g. the case of coulombic scattering) the hole transition rate has to be larger at resonance than the electron one. This holds only at resonance where for both electrons and holes the initial and final states are delocalised over the whole heterostructure. On the other hand, off resonance the spatial localisation of the initial and final states in different wells becomes much more pronounced for holes than for electrons when the energy detuning between the initial and final states increases, thereby decreasing sharply the transition rates. Thus, by varying F around a resonance one should find narrower resonance curves for the assisted transfer of holes than for that of the electrons if it were not for inhomogeneous broadenings. Actual quantum well samples display layer width fluctuations (islands). The sample can in the limit of large islands be envisioned as the superposition of many micro - samples, each characterised by different resonance fields. It is clear that the individual narrow hole resonances are more liable to distortions by inhomogeneous broadening effects than the relatively broad electron ones.

Let $\delta F$ be the width of an individual resonance occuring at $F_0$ and let $\Delta F$ be that of the resonance field distribution around $F_1$ due to the fluctuations in the layer width:

$$e\Delta F = | \, dE_1/dL \, | \Delta L \, / \, [h + (L + L')/2] \qquad (20)$$

where L and L' (h) are the average thicknesses of the two quantum wells (barrier) and where we have assumed for simplicity that only one well width (L) , and thus only one energy level ($E_1$), experiences fluctuations. Let us also empirically assume that the two functions describing the resonance versus $F(\tau^{-1}(F))$ and the resonance fields distribution ($P(F_0)$) are gaussians:

$$1/\tau(F) = 1/\tau_b + (1/\tau_0 - 1/\tau_b) \exp[-(F - F_0)^2/2(\delta F)^2] \qquad (21)$$

$$P(F_0) = [\Delta F \sqrt{2\pi}]^{-1} \exp[-(F_0 - F_1)^2/2(\Delta F)^2] \qquad (22)$$

where $\tau_0$ and $\tau_b$ are the transfer times at resonance and far away from resonance respectively for a given resonance field $F_0$. If we assume these two times to be independent of $F_0$ the resulting resonance curve $<1/\tau(F)>$ is the convolution of the two functions:

$$\langle 1/\tau(F)\rangle = 1/\tau_b + (1/\tau_0 - 1/\tau_b)\,(\delta F/\sigma F)\,\exp\left[-(F - F_1)^2/2(\sigma F)^2\right] \quad (23)$$

where:

$$\sigma F = \left[\,(\delta F)^2 + (\Delta F)^2\right]^{1/2} \qquad\qquad\qquad (24)$$

More detailed numerical simulations have been undertaken to evaluate the effect of inhomogeneous broadening. We have found that the electronic resonances are relatively unaffected by fluctuations of one monolayer while the hole ones are more changed , as expected. The inhomogeneous broadening effects are more important in heterostructures containing narrow wells (because the $|dE_1/dL|$ are larger in narrow wells) or narrow barriers (because the matrix elements appearing in the assisted transfer rates depend exponentially on the barrier thickness and thus fluctuates widely in the case of moderately transparent barriers). Note that the heterostructures with narrow wells and/or barriers are exactly the ones used to detect the hole transfer.

Although deep and sharp hole resonances can be washed out by inhomogeneous broadening effects (and also by homogeneous ones), the background hole transfer time is a quantity which is very little affected by the layer fluctuations. It is remarkable that off resonance the assisted hole transfer time can be found in the range 100-500 ps for reasonable amount of defects. This means that the very idea of "immobile" holes is incorrect in truly quantum heterostructures such as double quantum wells.

Some words of caution have to be added regarding the order of magnitude of the assisted transfer times: in our model (Born approximation) the transfer time $\tau$ is inversely proportional to the areal concentration of defects. $\tau$ also depends upon the material under consideration, for instance upon the valence band offset $|Vp|$. Since the transfer is very much favored by the band mixing effects there exist situations where the band mixing in the final states is small. This is for example realised when in a double well under appropriate bias the ground hole level of the thick well is made to anti-cross the hole levels predominantly localised in a very narrow well. A long transfer time is predicted to occur under these circumstances. The assisted transfer from the ground hole state of the narrow well to the hole states of the wide well should be faster, because of the larger band mixing effects in the final states on the one hand and of the larger density of final states on the other hand.

Recent time-resolved measurements of the photoluminescence of double quantum wells have evidenced the hole escape outside a wide quantum well. Alexander et al [16] have reported strong evidences for resonances in the hole transfer time versus the electric field in biased 50Å- h -100Å GaAs-$Ga_{0.65}Al_{0.35}As$ double quantum well structures and found short hole transfer times (some smaller than 20ps). At resonance the hole transfer time was found to increase with increasing intermediate barrier thickness h while off resonance (transfer time of the order of a few hundreds picoseconds) the trend was not so clear. In any event, the hole transfer time has been demonstrated to be quite short, i.e. shorter than or equal to the recombination lifetime.

Qualitatively similar results on the fast hole escape have been reported by Vodjani et al [17] in their study of the photoluminescence associated with the electron-hole recombination in the central well of an operating double barrier structure. In these experiments, under resonant tunnelling of electrons, the photoluminescence intensity is directly proportional to the number of minority holes injected in the well. Under illumination of the contact regions, at a smaller energy than the bandgap of the quantum well, the holes which are available for the recombination are those which have transferred from the contacts to the well. Thus, the photoluminescence intensity is a good indicator of the hole transfer.

## ACKNOWLEDGEMENTS

We thank the CAPES ( Brazil: R. F.) and CNRS ( France: G. B.) for financial support.

## REFERENCES

1)    M.Altarelli in *Semiconductor Superlattices and Heterojunctions* (Springer Verlag, Berlin 1986)

2)    G.Bastard, *Wave mechanics applied to semiconductor heterostructures* (Les Editions de Physique, Les Ulis, 1988)

3)    L.J.Sham, Superl. and Microstr. **5**,335 (1989)

4)    J. M. Luttinger, Phys. Rev. **102**, 1030 (1956)

5)    A. Twardowski and C.Hermann, Phys. Rev. B **35**, 8144 (1987)

6)    H.Chu and Y.C.Chang, Phys. Rev. B **39**, 10861 (1989)

7)   R.Ferreira and G.Bastard, submitted to Phys.Rev.B  (1990)

8)   T.Uenoyama and L.J.Sham, Phys.Rev.Lett. **64**,3070 (1990)

9)   G. Bastard, J. A. Brum and R. Ferreira (1989), to be published in Solid State Physics

10)  G. H. Wannier, Rev. Mod. Phys. **34**, 645 (1962)

11)  R.Tsu and G. Döhler, Phys.Rev.B **12**,680 (1975)

12)  J. Bleuse and P. Voisin, 1988 (unpublished)

13)  T. Weil and B. Vinter, App. Phys. Lett. **50**, 1281 (1987) see also B. Vinter and F. Chevoir, to be published in the Proceedings of the NATO Workshop on Resonant Tunnelling. El Escorial .(1990)

14)  R.Ferreira and G. Bastard, Europhys.Lett. **10**,279 (1989)

15)  K. Wassel and M. Altarelli, Phys. Rev. **B39**, 12803 (1989)

16)  M. G. W. Alexander, M. Nido, W. W. Rühle and K. Köhler, to be published in the Proceedings of the NATO Workshop on Resonant Tunnelling. El Escorial (1990)

17)  N. Vodjani, D. Côte, F. Chevoir, D. Thomas, E. Costard, J. Nagle and P. Bois, Semicond. Sci. Technol. **5**, 538 (1990)

# IMPURITIES IN SEMICONDUCTORS

H. G. Grimmeiss, M. Kleverman, J. Olajos, P. Omling
and V. Nagesh

Department of Solid State Physics
University of Lund, P.O.Box 118
S-221 00 Lund, Sweden

A brief outline is presented on recent developments in defect characterization
and identification in semiconductors which have been made possible by the application of methods other than junction space charge techniques (JSCT). Chalcogens
and several transition metals in silicon are used as examples in order to show how
important parameters and properties of defects can be revealed by using spectroscopic methods. One of the methods, namely photothermal ionization spectroscopy
is discussed in more detail. Si/Ge is taken as an example to show how JSCT can be
used for the study of low-dimensional structures.

## INTRODUCTION

The great progress in semiconductor electronics can be traced to a unique combination of basic conceptual advances, the perfection of new materials and the development of new device principles. Ever since the invention of the transistor, we
have witnessed a fantastic growth in silicon technology, leading to more complex
functions and higher densities of devices such as the mega bit memories. This development would hardly be plausible without an increased understanding of semiconductor materials and a better insight into the important role, defects play in
most currently used devices. It is therefore not surprising that a variety of techniques for the characterization and identification of defects in semiconductors have
evolved during the last few decades. Some of the earlier methods comprise electrical and magnetic measurements, photoconductivity, absorption and various forms
of luminescence [1]. All these methods allowed the study of thermal "activation energies" and/or optical threshold energies but only very rarely could absolute values of fundamental electronic parameters be determined [2]. This situation has

changed considerably with the introduction of junction space charge techniques (JSCT) such as photocurrent [3] and dark capacitance [4] measurements. Most of these methods, in particular deep level transient spectroscopy (DLTS) [5], allows the determination of defect concentrations with sufficient accuracy and the direct measurement of electronic parameters such as emission and capture rates. Once these parameters are known, the energy position of defects within the band gap can be determined in terms of enthalpies and/or threshold energies deduced from the distribution of a single optical cross section. There is no doubt that these measurement methods provided an important breakthrough in the characterization of defects in semiconductors.

In parallel with this development, fundamental improvements were achieved in defect identification. In this paper, for the sake of simplicity, we will use EPR (electron paramagnetic resonance) as a synonym for defect identification. Application of EPR to semiconductors produced another breakthrough [6], this time in the area of defect *identification*. It is, however, important to note that results obtained from EPR, quite often cannot be directly related to a defect which has been studied by JSCT. This implied that many defects in semiconductors which have been accurately characterized have never been identified and that quite a number of defects which have been identified by EPR have not been characterized in detail since EPR is not the perfect tool for studying electronic parameters. Techniques combining spin-dependent data with those from characterization methods are therefore extremely important. Photo-EPR is one such example which is extremely powerful in correlating EPR data with those from absorption or JSCT [7].

The object of this paper is to briefly outline recent developments in deep-level characterization and identification which have been made possible by the application of methods other than JSCT. Besides EPR, we will in particular focus our attention on two methods, namely absorption and photothermal ionization spectroscopy (PTIS) which have proven to be powerful tools in characterization of many defects [8]. PTIS and absorption possess many common features but nevertheless have interesting differences. Furthermore, in contrast to luminescence measurements, both PTIS and absorption measurements can be used in the study of defects, regardless of whether they are considered as radiative or non-radiative recombination centres.

PHOTOTHERMAL IONIZATION SPECTROSCOPY

For more than 20 years photothermal ionization spectroscopy has been used in the study of shallow centers. Only recently the method has been applied to investigate deep centers. Previous investigations of shallow centers have shown that under certain conditions, especially in a particular temperature range, the photoconductivity spectrum is a line spectrum reflecting the energy level structure of the centres. In contrast to absorption measurements, however, the PTIS signal has been shown to be independent of the concentration of defects down to very low values [8]. PTIS is therefore particularly suited for the study of deep centers since the solubility of impurities often decreases with increasing binding energies [9].

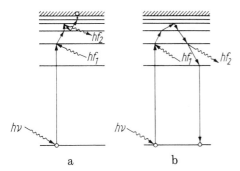

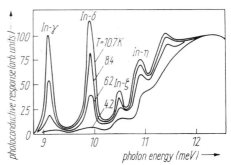

Fig. 1 Scheme of the thermal transitions in an impurity atom excited by light. (a) Ionization (b) return to the ground state. hv energy of light quantum, hf phonon energy [8].

Fig. 2 PTIS spectra of indium-doped germanium (p-type) at different temperatures [8].

The PTIS signal arises as a result of a two-step excitation process consisting of an optical excitation from the ground state into excited states followed by a thermal ionization of the excited defect (Fig. 1). It should be noted that the optical excitation and, hence, the absorption of photons may or may not result in a positive PTIS signal depending on whether or not the charge carrier is finally excited into the band continuum. Fig. 2 shows typical spectra obtained already in 1967 for indium-doped germanium at different temperatures [8]. Indium has a binding energy of 11.7 meV in germanium. Transitions from the ground state into the valence band continuum are clearly seen as a broad, featureless spectrum at energies greater than the binding energy. The five lines on the low-energy side of the continuum are caused by the two-step excitation process via excited states. Due to the two-step excitation process the peak height of the lines increases with increasing temperature implying that the signal of shallower levels is favoured over the signal from deeper levels, as seen in the spectrum taken at the lowest temperature.

DEFECT CHARACTERIZATION

A quantitative and qualitative description of a deep level defect in a semiconductor would be incomplete without the following parameters or properties: (1) Concentrations (2) Thermal and optical emission and capture rates (3) Electronic structure (4) Charge state (5) Whether it is an isolated defect or not (6) Defect symmetry (7) Lattice relaxation (8) Type of defect, i.e. whether it is a donor or an acceptor and (9) Chemical identity. Depending on the defect there may be other parameters or properties such as metastability etc. which are of interest but due to the limited scope of this paper we will restrict ourselves to the list above.

Concentration of the defect and its emission and capture rates are best studied by JSCT and will not be further discussed in this paper. In contrast to methods such as PTIS, absorption or luminescence, JSCT cannot be considered as truly spec-

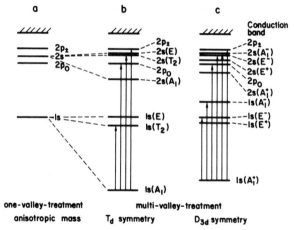

Fig. 3 The lowest-energy states of an electron bound to a donor in silicon: (a) where one-valley-treatment is used; (b) when multi-valley treatment and $T_d$ symmetry are used; (c) when multi-valley treatment and $D_{3d}$ symmetry are used. Allowed optical no-phonon transitions are marked with arrows.

troscopic techniques, solely on the basis of resolution. JSCT are limited by about 10 meV resolution, whereas other ("true") spectroscopic methods have a resolution of about 10 μeV or even better corresponding to an increase in resolution by three orders of magnitude. The enrgy position of a defect can therefore in general only be obtained within 10 meV by JSCT which is not sufficient for studying the electronic structure of defects.

ENERGY LEVEL STRUCTURE OF SIMPLE DEFECTS

Theoretically, the energy structure of defects is well formulated by effective mass theory [10] (EMT) and is experimentally investigated by absorption measurements. In silicon for example EMT considers a substitutional group V donor as a hydrogen-like atom with a screened Coulomb potential which results not only in an energy level in the band gap for the donor ground state but also in a series of excited states (Fig. 3). Incorporating the anisotropic effective mass tensor of the conduction band electrons in EMT calculations, a splitting of p-like states into a $p_o$ state (m=0) and a $p_\pm$ state (m=±1) is obtained. Agreement with experimental results for the enrgy level structure of for example group V donors in silicon is very good except for s states. The deviation from s states in such a simple EMT treatment is caused by the multivalley nature of the conduction band which partly lifts the 6 fold degeneracy (excluding spin) of the s states. It can be shown that when a donor occupies a substitutional site ($T_d$ symmetry), an s state splits into a singlet $A_1$ state, a triplet $T_2$ state and a doublet E state [10] (Fig. 3 b). If the symmetry of the defect is not $T_d$, the splitting of the s states will be different, as shown in Fig. 3 c. Using the electronic structure obtained from EMT together with selection rules for transitions from the ground state to excited states it is in principle possible to predict the absorption spectrum (arrows in Fig. 3) of for example group V and VI impurities in silicon. As shown in Fig. 4 the agreement between EMT and experiment is excellent for transitions to p-like states and well within the experimental error.

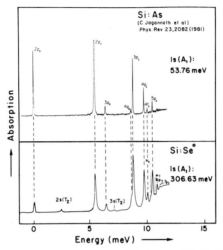

Fig. 4  Line spectrum of Si:As and Si:Se⁰.

CHARGE STATES OF DEFECTS

It has been shown previously that group VI elements form double donors in silicon [7,11-14]. In the neutral state these donors bind two extra electrons and therefore have three different charge states. It is well known that many transition metals (TM) in silicon can also have different charge states resulting in donor- and/or acceptor-like excitation spectra [15]. Since emission and capture rates are often strongly dependent on the type of defect, knowledge of the charge state of defects is desirable.

Once the line spectrum of a defect is obtained and correctly analyzed it is rather trivial to identify the charge state of the defect studied [14] (as shown in Fig. 5). Since the spacing of p-like excited states is proportional to $Z^2$ where Z is the charge of the defect, Z is directly obtained from the analysis of the spectra. Determining the charge state is not only interesting per se but often very helpful in correlating for example, different peaks observed in DLTS [16].

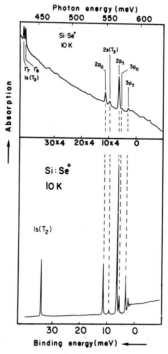

Fig. 5   Line spectrum of Si:Se⁺ (a) compared with the absorption spectrum of isolated neutral selenium in silicon (b).

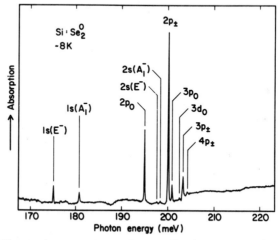

Fig. 6   Absorption spectrum of neutral selenium pairs in silicon.

# ISOLATED DEFECTS OR COMPLEXES?

Information on the microscopic structure of a defect is normally obtained from spin-dependent measurements such as EPR or ENDOR (electron nuclear double resonance). However, some valuable insight into the symmetry of a defect can also be obtained from spectroscopic methods such as absorption or PTIS [14]. As previously mentioned the splitting of s-states due to valley-orbit interaction depends on the symmetry of the defect and therefore in certain cases indicates whether the spectrum originates from an isolated defect or a complex. In this respect, chalcogens in silicon and germanium are ideal examples, since the analysis is facilitated by the similarity of the chalcogen spectra with those originating from group V donors. The fact that only one 1 s line due to the transition $1s(A_1)$-$1s(T_2)$ is observed in Fig. 5 for Si:Se suggests that the spectrum is generated by an isolated defect with $T_d$ symmetry in agreement with Fig. 3 b and EPR data [6,7] as well as theoretical considerations [17]. In the same sample additional spectra due to other selenium related centers are also observed. One example is shown in Fig. 6 which according to Fig. 3 should have $D_{3d}$ symmetry (two 1 s lines) and has been proven to be due to neutral selenium pairs [14]. These examples show that line spectra may indeed give valuable information on the symmetry of a defect and, hence, on its microstructure. A number of centres have already been analyzed in this manner and useful information on their microstructures have been obtained.


# LATTICE RELAXATION

The discussion thus far has been restricted to donors. Without going into details it is worth mentioning that similar line spectra have also been observed for deep acceptors in both silicon and germanium. Of particular interest in this context are defects due to transition metals such as Au, Pt, Ag and Mn [18-22]. Some of them are fast diffusers and play an important role in manufacturing microelectronic circuits. Furthermore, several of these transition metals are known as lifetime killers already when present in small concentrations. Their presence is therefore highly undesirable in certain devices. A complete understanding of their electronic properties and energy level structure as well as characteristic fingerprints for identification and simple experimental testing is thus in many respects considered as useful.

Fig. 7 compares the line spectra obtained for indium, gold and platinum in silicon. The striking similarity of the Au and Pt spectra with the group III acceptor In is rather obvious. However, there are two features which are worth mentioning. In addition to the $P_{3/2}$ and $P_{1/2}$ line series, all three spectra exhibit a detailed structure at higher energies. Later this structure will be shown to be due to phonon-assisted Fano resonances [23]. Furthermore, a closer inspection of the Au and Pt spectra show [20] that the Pt spectrum consists of additional lines which have been identified as phonon replicas of the corresponding $P_{3/2}$ and $P_{1/2}$ line series (Fig. 8). The phonon involved is resonant with the bulk accoustic phonons and is best discussed in terms of a pseudo-localized phonon of about 7 meV which has an appreciable amplitude only in the vicinity of the impurity. A Huang-Rhys factor [24] of about 0.4 was estimated for the Pt-center which indicates a very weak electron-pho-

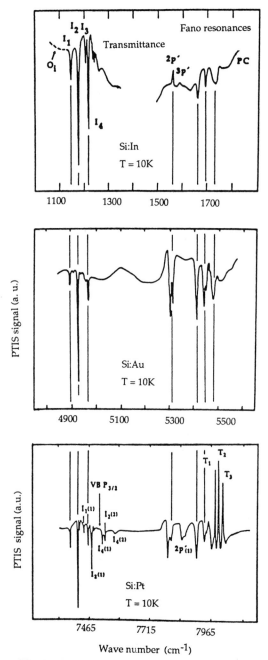

Fig. 7 Comparison of spectra obtained for different acceptors in silicon.

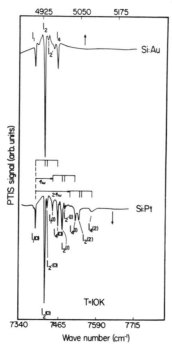

Fig. 8  Comparison of the Au and Pt line spectra.

non coupling and a Franck-Condon shift of less than 3 meV. This value is a direct measure of lattice relaxation which in the case of platinum is rather small.

ACCEPTOR OR DONOR LEVEL?

It has already been mentioned that several transition metals form defects with at least three different charge states involving both acceptor- and donor-like properties. From what has been said so far the type of center should be easily deduced from the characteristic features of optical excitation spectra. This is certainly true for simple line spectra such as those for isolated chalcogens or gold in silicon. It will be shown later, however, that due to variations in selection rules or more complex structures of the initial state, a great number of defects may generate rather complicated line spectra which cannot always be analyzed in a straight-forward manner. Additional information on the type of center is then very useful not only for purposes of characterization but also in the analysis of the spectra.

All defects hitherto discussed show associated resonance structures in the continuum part of the spectra which are caused by a higher order excitation process. A

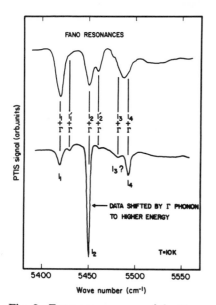

Fig. 9 Fano resonances of the $P_{3/2}$ excited states for the gold acceptor in silicon.

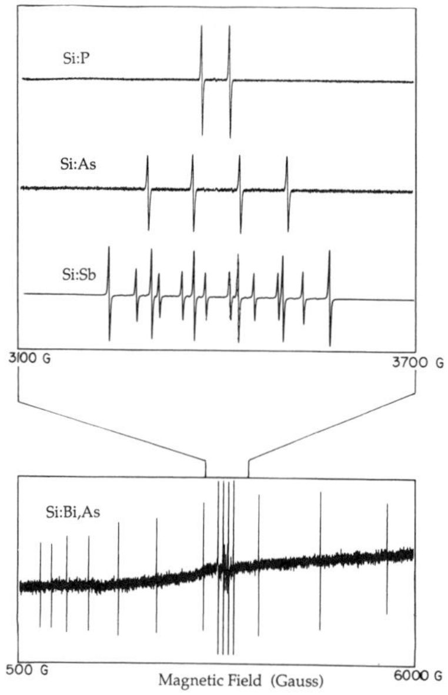

Fig. 10 EPR spectra of three group V donors in silicon [27].

bound electron together with a phonon may interact resonantly with the electronic continuum. Due to stringent selection rules the resonant electron phonon state will be quasi-discrete leading to the rather sharp features observed in Fig. 7 which have been denoted as phonon-assisted Fano resonances [23,25]. For donors in silicon, only intervalley phonons are involved in the resonances [23,26], whereas for acceptors like the isolated gold the optical zone-center phonon $\Gamma$ participates in the process [25], as shown in Fig. 9. Hence, by determining the phonon participating in the Fano resonances it is possible to decide whether the acceptor- or donor-like version of a defect is being studied.

IDENTIFICATION

Spin dependent measurements are still by far the best techniques for identifying defects in semiconductors. The resonances observed are often a direct fingerprint of the chemical species involved in the defect. This is illustrated in Fig. 10 by the EPR spectra of three group V donors in silicon [27]. The number of equidistant lines observed are related to the nuclear spin I and given by 2I+1.

No known isotopes of phosphorous or arsenic exsist, hence, only two and four lines, respectively, are observed. The natural abundance of antimony is 57.3% $^{121}$Sb (I=5/2) and 42.7% $^{123}$Sb (I=7/2) and therefore a total of 6+8 lines are observed. To-

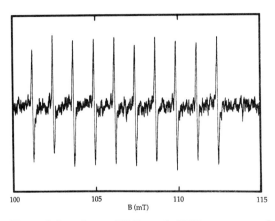

Fig. 11 Part of the trigonal FeIn-pair EPR spectrum showing the hyperfine structure due to [115]In and [113]In [28].

gether with the angular dependence and other measurements the symmetry and chemical identity of a defect can be revealed from such studies. This is in particular very useful if the EPR active center is not an isolated defect but is more complex in nature. Fig. 11 shows part of the trigonal FeIn-pair spectrum with the magnetic field parallel to the trigonal axis and featuring the hyperfine structure due to [115]In ($I=9/2$, 95.72% natural abundance), and [113]In ($I=9/2$, 4.28% natural abundance) [28]. This manifests the presence of In in the defect but gives no proof of the presence of Fe, since normal Fe contains only a very small amount of an isotope with $I \neq 0$ and therefore does not contribute to the hyperfine splitting of the spectrum. The involvement of Fe in the defect is studied by using iron enriched to 34.6%

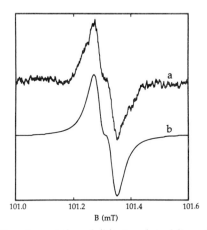

Fig. 12 (a) Exprimental and (b) simulated line shape of the spectrum shown in Fig. 11 exhibiting the additional hyperfine structure due to [57]Fe ($I=1/2$, enriched to 94.6% abundance) [28].

27

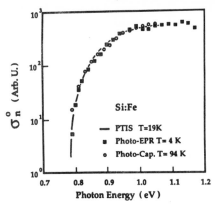

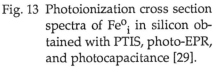

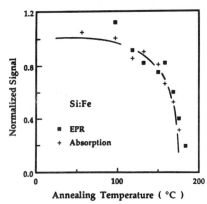

Fig. 13 Photoionization cross section spectra of $Fe^o_i$ in silicon obtained with PTIS, photo-EPR, and photocapacitance [29].

Fig. 14 $Fe^o_i$-related EPR signal shown together with the intensity of the absorption lines, as a function of isochronical annealing temperature [29].

abundance with $^{57}Fe$ which has $I=1/2$, thus causing a splitting of each of the ten lines shown in Fig. 11 (Fig. 12).

Although the microstructure of the trigonal FeIn pair has been established and the chemical constituents identified, we have not yet succeeded in determining the various electronic properties of this defect. Connecting EPR data with the energy structure and electronic parameters obtained from other measurements has, nevertheless, been demonstrated for several other defects in silicon and makes the detailed characterization finally meaningful. In this paper we will illustrate such a correlation with a simple example, namely the interstitial neutral $Fe_i$ center in silicon [29]. By measuring the same physical parameter with EPR and at least one of the other characterization methods it is possible to correlate electronic parameters with chemically identified defects. In the case of $Fe_i^o$ we used the photoionization cross section of electrons $\sigma_n^o$ and the annealing properties of the center to demonstrate that the electronic parameters obtained are in fact related to the $Fe_i^o$ center. The EPR signal was therefore studied by illuminating the sample in situ with monochromatic light of variable wavelength and measuring the change in the charge state. The spectrum of $\sigma_n^o$ was then compared with those from PTIS and photocapacitance (Fig. 13). The good agreement between the three spectra strongly suggests that the line spectrum and electronic parameters observed in Si:Fe do indeed originate from the $Fe_i^o$ center. Further confirmation was obtained from annealing experiments in which the $Fe_i^o$-related EPR signal was compared with the intensity of the absorption lines as a function of the isochronal annealing temperature (Fig. 14).

The discussion thus far, has been restricted to fairly simple optical spectra. Such defects are rather the exceptions. In general more complicated spectra are observed. The $Fe_i^0$ center mentioned in the previous section is a typical example. The defect exhibits a transmission spectrum consisting of three overlapping EMT-donor series as shown in Fig. 15. Although the positions and intensities of the p-like lines are in good agreement with those predicted by EMT, no sharp line transitions to valley-orbit split 1s states were detected even though the spectrum includes 2s and 3s lines [29]. This implies that details of the symmetry of the center are difficult to obtain from such spectra without applying perturbation spectroscopy.

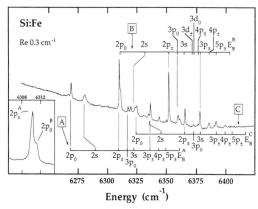

Fig. 15   High-resolution spectrum of $Fe_i^0$ in silicon [29].

Another example is the neutral interstitial manganese center in silicon [22] which has previously been studied by EPR, Hall-effect measurements and JSCT [30-32]. These investigations showed that the defect has three different charge states ($Mn^-$, $Mn^0$ and $Mn^+$) and that the energy position of the acceptor-like version is about $E_c$-0.13 eV corresponding to the transition from $Mn^-$ to $Mn^0$. PTIS and transmission measurements of manganese diffused at 1250 °C and rapidly quenched samples, revealed [22] a series of four sharp lines at 2K (Fig. 16). The energy between the lines decreased with increasing photon energy indicating that the lines originate from excited states due to a screened Coulomb potential. From the high-energy limit of the line series a binding energy at 1048 meV was deduced which is in reasonable agreement with the previously suggested position of $E_c$-0.13 eV for the Mn acceptor considering that the band gap energy of silicon is about 1170 meV at low temperatures and assuming small lattice relaxations. Since the

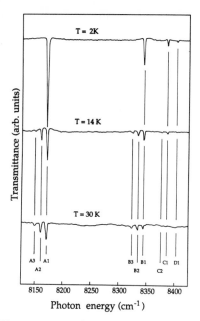

Fig. 16    Line spectra of $Mn^0_i$ in silicon at different
temperatures [22].

$Mn_i^0$ and $Mn_i^-$ EPR signal did not change with illumination, photo-EPR could not be applied and information on the chemical identity of the defect was therefore only obtained from an isochronal annealing experiment. All lines of the $Mn_i^0$ spectrum showed the same annealing behaviour. A comparison of the total transmission intensity with the $Mn_i^0$ integrated EPR signal (Fig. 17) as a function of the annealing temperature strongly suggests that the optical spectra originate from the $Mn_i^0$ center. Since the $Mn_i^0$ and $Mn_i^-$ EPR signals showed identical annealing behaviour it is reasonable to assume that no shift of the Fermi level position occurred during the annealing experiment.

When raising the measurement temperature the intensity of the transmission lines decreased and new lines were observed at slightly lower energies (Fig. 16). The energy distance between the lines in each group is almost identical for all groups revealing a mean value of 1.3 meV. This and the observation that the relative intensities of the lines in each group are similar, suggests that the splitting has a common origin and may be caused by the electronic or vibronic structure of the initial or final state. Based on a detailed analysis it has been suggested that the splitting of the lines is due to a splitting of the initial state in the optical transitions and caused by a local phonon in agreement with the isochronal annealing experiment (Fig. 18). It should be noted that the energy of the local phonon is much smaller than those observed previously for TM related impurities in silicon corresponding to a much smaller spring constant.

The transition model for the observed line spectrum of $Mn_i^0$ (Fig. 18) fits reasonably well into the trend found for the group III acceptors in silicon if it is assumed that the line with the lowest energy originates from transitions to the second s state ($2\Gamma_8^+$). The energy levels of the even parity s states have been determined

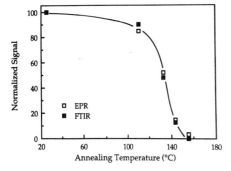

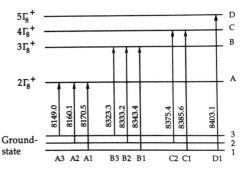

Fig. 17 Comparison of the total transmission intensity with the Mn$^O_i$ integrated EPR signal as a function of annealing temperature [22].

Fig. 18 Transition model for the observed lines shown in Fig. 16 (all energies given in cm$^{-1}$) [22].

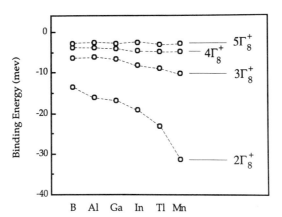

Fig. 19 Energy level model of even parity s states for different acceptors in silicon [22].

experimentally for group III acceptors. They all show a considerable central-cell shift which increases with the corresponding increase in ground state binding energy (Fig. 19). In the case of $Mn_i^0$ it was therefore reasonable to assign the final hole states as different excited s states in agreement with the relative intensities of the lines and previous results obtained for the Ag-related donor in silicon [21].

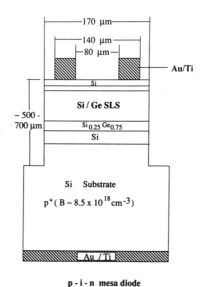

**p - i - n mesa diode**

Fig. 20   Diode structure.

## DEFECTS IN LOW-DIMENSIONAL STRUCTURES

JSCT have successfully been applied to diodes with space charge regions consisting of a superlattice. The diode structure which will be discussed here is shown in Fig. 20. It is worth mentioning that the sample which is basically a $p^+n$ Si/Ge diode had excellent I-V characteristics with reverse currents smaller than 10-13 A at 77 K for voltages less than 1V. The spectral distribution of the photoionization cross section for electrons obtained from transient photocapacitance measurements [4,7] is shown in Fig. 21. There are good reasons to believe that the signal almost entirely originates from the superlattice. Two features of the spectrum are of particular interest: (1) The optical cross section does not increase smoothly with photon energy but exhibits a step-like behaviour at certain photon energies. (2) The threshold energy of the spectrum is about 0.35 eV which is much smaller than the expected band gap of the superlattice at about 0.8-0.9 eV. This suggests that the observed effect is due to localized defects rather than transitions between bands. If it is

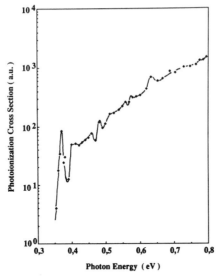

Fig. 21    Spectrum of photoionization cross section.

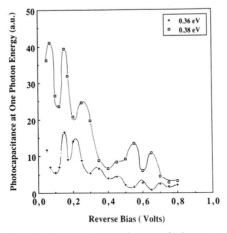

Fig. 22   Electric field dependence of photoionization
cross section for two different photon energies.

assumed that the steps seen in the spectrum of Fig. 21 are caused by the Wannier-Stark effect one would expect that the photoionization cross section at a fixed energy inside the F=0 miniband oscillates with the electric field F [33] and that the peaks of these oscillations shift with changing photon energy. This is indeed observed (Fig. 22). Similar studies have previously been performed in AlGaAs [34] but to the best of our knowledge these are the first investigations in Si/Ge [35].

## CONCLUSION

The intention of this short review was to show that conclusive evidence concerning the identification and characterization of defects in semiconductors is normally only obtained from a combination of different techniques. Recent developments in high resolution spectroscopy have provided complementary methods for a more detailed analysis of defects. Additional information can be obtained from these spectroscopic methods by applying perturbation spectroscopy such as Zeeman or piezo spectroscopy. However, due to the limited scope of this paper it was not possible to discuss these possibilities in more detail.

## REFERENCES

[1] See for example R. H. Bube, *Photoconductivity of Solids*, John Wiley & Sons, Inc., New York, 1960
[2] See for example W. Hoogenstraaten, Thesis, University of Amsterdam (1958)
[3] G. Björklund and H. G. Grimmeiss, Phys. Status Soldidi **42**, K1 (1970)
[4] C. T. Sah, L. Forbes, L. L. Rosier and A. F. Tasch, Solid State Electron. **13**, 759 (1970)
[5] D. V. Lang, J. Appl. Phys. **45**, 3014 and 3023 (1974)
[6] See for example G. W. Ludwig, Phys. Rev. **137**, A1520 (1965)
[7] H. G. Grimmeiss, E. Janzén, H. Ennen, O. Schirmer, J. Schneider, R. Wörner, C. Holm, E. Sirtl and P. Wagner, Phys. Rev. **B24**, 4571 (1981)
[8] T. M. Lifshits, N. P. Likhtman and V. I. Sidorov, Fiz. Tekh. Poluprov. **2**, 782 (1968), Sh. M. Kogan and T. M. Lifshits, Phys. Status Solidi (a) **39**, 11 (1977) and references therein
[9] H. J. Queisser, Festkörperprobleme **XI**, 45 (1971)
[10] W. Kohn and J. M. Luttinger, Phys. Rev. **98**, 915 (1955)
[11] W. E. Krag and H.J. Zeiger, Phys. Rev. Lett. **8**, 485 (1962)
[12] W. E. Krag, W. H. Kleiner and H. J. Zeiger, Phys. Rev. **B33**, 8304 (1986)
[13] H. G. Grimmeiss and E. Janzén, in Defects in Semiconductors, Proc. of the Material Research Society's Annual Meeting, 1982
[14] E. Janzén, R. Stedman, G. Grossmann and H. G. Grimmeiss, Phys. Rev. **28**, 1907 (1984)
[15] A. Zunger, Solid State Physics, **39**, 275 (1987)
[16] H. G. Grimmeiss, L. Montelius, and K. Larsson, Phys. Rev. **37**, 6916 (1988)
[17] M. Scheffler, F. Beeler, O. Jepsen, O. Gunnarsson, O.K. Andersen and C.B. Bachelet, J. Electr. Materials, December 1984
[18] M. Kleverman, J. Olajos, and H. G. Grimmeiss, Phys. Rev. **35**, 4093 (1987)
[19] G. Armelles, J. Barrau, M. Brousseau, B. Pajot, and C. Naud, Solid State Commun. **56**, 303 (1985)

[20] M. Kleverman, J. Olajos and H. G. Grimmeiss Phys. Rev. **B37**, 2613 (1988)

[21] J. Olajos, M. Kleverman and H. G. Grimmeiss, Phys. Rev. **B38**, 10633 (1988)

[22] T. Bever, P. Emanuelsson, M. Kleverman, and H. G. Grimmeiss, J. Appl. Phys. Lett. **55**, 2541 (1989)

[23] E. Janzén, G. Grossman, R. Stedman, and H. G. Grimmeiss, Phys. Rev. **31**, 8000 (1985)

[24] K. Huang and A. Rhys, Proc. Roy. Soc. **204**, 406 (1950)

[25] G. D. Watkins and W. B. Fowler, Phys. Rev. **B16**, 4524 (1977)

[26] W. Harrison, Phys. Rev. **104**, 1281 (1956)

[27] P. Omling, private communication

[28] P. Omling, P. Emanuelsson, W. Gehlhoff and H. G. Grimmeiss, Solid State Commun. **70**, 807 (1989)

[29] J. Olajos, B. Bech Nielsen, M. Kleverman, P. Omling, P. Emanuelsson, and H. G. Grimmeiss, Appl. Phys. Lett. **53**, 2507 (1988)

[30] G. W. Ludwig and H. H. Woodbury, Solid State Phys. **13**, 223 (1962)

[31] H. Feichtinger and R. Czaputa, Appl. Phys. Lett. **39**, 706 (1981)

[32] R. Czaputa, H. Feichtinger and J. Oswald, Solid State Comm. **47**, 4, 223 (1983)

[33] J. Bleuse, G. Bastard and P. Voisin, Phys.Rev.Lett. **60**, 220 (1988)

[34] E. E Mendez, F. Agulló-Rueda. and J. M. Hong, Phys.Rev.Lett. **60**, 2426 (1988)

[35] H. Presting, H. G. Grimmeiss, V. Nagesh, H. Kibbel and E. Kasper, Proc. of the 20th ICPS, Thessaloniki, 1990

# PHOTOLUMINESCENCE OF 3D AND LOW DIMENSIONAL SYSTEMS

B. Hamilton

Department of Pure and Applied Physics
University of Manchester Institute of Science and Technology
P.O. Box 88, Manchester M60 1QD, U.K.

## INTRODUCTION

This paper deals with various topics in the field of photoluminescence, and is meant to set the scene for later papers which deal with rather specialised reports of low dimensional systems. An important aspect of what is written below concerns experimental techniques, and the way in which advances in techniques allow us to probe further the fine detail of complex semiconductor structures. The literature on photoluminescence is vast, and so I have tried to emphasise basic mechanisms and to illustrate how photoluminescence spectroscopy has evolved to explore these mechanisms.

## THE BASIC FEATURES OF PHOTOLUMINESCENCE

The simplest and most common way to perform photoluminescence is to cool the structure to be measured down to some rather low temperature (typically 4.2K), and to excite it with a laser source with photon energy greater than the forbidden gap. The laser energy is absorbed, principally by excitation of electrons from the valence band to the conduction band. Some of this input energy is converted to lower energy photons by radiative recombination; this constitutes the photoluminescence signal, which is detected and analysed. In fact the detailed sequence of events can be extremely complicated. A useful way to view this sequence is to consider the energy migration steps, starting with the initial optical absorbtion process and ending with a radiative decay channel.

The common choice of laser photon energy to be safely above the energy gap, should be thought of as a non resonant excitation condition in the sense that it is not tuned to any particular optical transition. It will be shown below that this choice, although simple from the experimental viewpoint, often results in rather complex energy migration, and may hide much useful spectral information. The energy migration features resulting from non resident excitation are generally the same for 3-D and low dimensional systems, and are tabulated in figure 1. Each stage of the energy migration is characterised by some dynamical process (exciton formation, diffusion, capture etc.), and it is important to note that,**in principle**, photoluminescence measurements can provide information on all of these processes.

It is generally true that most photoluminescence experiments liquid helium temperatures aer used (<5K) in order to narrow line widths and to prevent the thermal quenching of spectral features. The use of low temperatures enhances the importance of excitons in the optical spectra, since in most semiconductors excitons are relatively thermally stable below 5K.

This is most dramatically revealed for the case of excitons bound to impurities in III-V materials; for three dimensional structures, impurities inevitably dominate the low temperature PL spectra, even in the purest systems grown to date. In low dimensional structures, where the exciton binding energy is greatly increased, spectra are may be dominated by the intrinsic exciton decay processes, particularly in the purest systems (GaAs based).

Absorption of the non-resonant laser line creates hot electrons and holes, high in their respective bands. In a direct energy gap semiconductor the electron wave vector **k**, is conserved without phonon participation, during a radiative decay process.

**NON RESONANT EXCITATION IN 3D SYSTEMS**

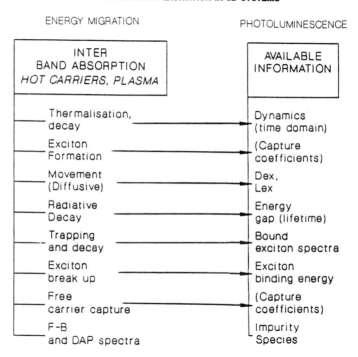

Figure 1.

Using figure 1 as a guide to what observable properties might be studied, it is clear that most published data concentrate on the spectral features characteristic of the last stages of the energy migration chain. This is a natural consequence of performing steady state experiments. At short times, typically of the order picoseconds or less, luminescence decay can be used to investigate carrier cooling rates and the electron phonon interactions which control cooling. Often such measurements are made with high excitation, and it is actually the properties of electron hole plasmas that are measured (1).

Moving down the energy migration path, the condensation of electrons and holes into excitons and the subsequent movement of excitons is to be expected. It should be noted that in heavily doped semiconductors, or highly defected systems, the excitons often become screened, and exciton effects are, at best, very weak. Also in indirect gap, high effective mass systems like Si and Ge, clusters of excitons (bi-excitons and tri-excitons), and even a completely new phase, "electron-hole liquids", can be observed (2). The formation and diffusion of excitons are relatively poorly understood phenomena. The direct movement of excitons is difficult to observe, and the exciton diffusion coefficients poorly understood, with some of the most reliable data available on Si (3).

In III-V materials there are several reports on exciton transport, both vertically (in the direction of crystal growth), (4), and in the plane of quantum well structures (5), (6). A complicating factor in such measurements is often the high excitation regimes needed to make the experiment work. At high excitation levels, the carrier gasses are, at least initially, likely to exist as plasmas, and the real temperature profiles during the measurement may be complex. It has been reported (5), that for the case of in plane exciton transport in GaAs/AlGaAs quantum wells, exciton transport can be described by isothermal diffusion influenced by interface scattering at the walls of the quantum well. The importance of phonon interactions in this materials system have been suggested (6), with transport being driven by a phonon wind, and transient confinement of carriers by phonon hot spots at very high carrier densities being used to interpret the data.

The radiative decay of the exciton is, of course, the process which gives the spectral output detected in steady state PL measurements. We should note that at high temperatures, when the exciton is dissociated, free carrier band to band recombination takes place. This intrinsic recombination is exploited in, say GaAs LED's and Lasers. The efficiency of this process is dictated by the trade off between radiative and non-radiative recombination in the system, and the measurement of PL decay is an important characterisation method for such device based materials (7). In quantum well and double heterojunction structures, the intrinsic luminescence decay can be an important guide to the strength of the non-radiative recombination taking place at the interfaces (8), (9).

The impurity related transitions dominate the low the low temperature of three dimensional III-V systems. The reason for this is that the exciton trapping processes are often very efficient at impurities; the free exciton is less thermally stable and is subject surface recombination. Although free exciton recombination can be seen in high purity GaAs and InP, it has never been shown to dominate the spectrum. There exists a vast literature on impurity related spectroscopy (10), and the importance of understanding such impurity and defect systems has led to very significant advances in spectroscopic techniques.

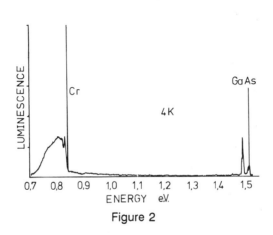

Figure 2

In fact the resonant excitation methods, briefly reviewed in the next section grew out of the need to improve information on impurities. The main problem underpinning non-resonant PL on impurity systems is encapsulated in figure 2. This is a spectrum of a high quality undoped GaAs epitaxial layer grown by MBE on a VG V90 machine. The layer (one micron thick) was grown onto a chromium doped semi-insulating substrate; the high quality of the layer allowed the rapid measurement of PL from both the layer and the substrate. In fact, the wide energy range shown in the figure was measured in around one second using Fourier Transform spectroscopy. The chromium gives rise to a deep level band, with a sharp no-phonon line near 0.84 eV and a well understood multiphonon side band (11). The GaAs impurity spectra due to exciton capture and decay at donors and (carbon) acceptors inadvertently present in the layer, are clearly distinguished in the spectrum. The donors are the highest energy feature, corresponding to the weakest exciton binding. Although acceptor species can be distinguished fairly readily donors cannot. Resonant excitation, and additionally, magnetic field application, enormously assists the spectroscopic analysis of both types of impurity.

## 3 RESONANT EXCITATION AND MAGNETIC FIELD ENHANCEMENT OF IMPURITY SPECTRA

The donor discrimination problem in II-V materials results essentially from the homogeneous line broadening   due to the random electric field distribution due to background concentration of charged impurities in the system. This Stark broadening mechanism combined with other mechanisms, leads to measured donor bound exciton line widths of, at best, 1 meV. Since the exciton binding energy of different donors is much less than this, all donors contribute to a unresolved "donor line"

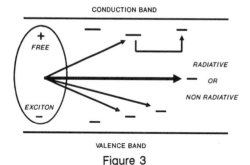

ENERGY MIGRATION BETWEEN SITES:
EFFICIENT CHANNELS ARE SELF SELECTING

Figure 3

In fact this behaviour is at the heart of a major shortcoming characteristic of non-resonant methods. The exciton is a mobile particle, as it diffuses through the crystal, it will be trapped with varying degrees of efficiency at many binding potentials, from which it may escape or at which it may decay. In other words the most efficient channels are self selecting, and the spectrum reflects this (fig. 3).

Resonant excitation within any broadened PL line effectively selects a subset of impurity levels; excitons are resonantly created at selected sites characterised by similar effective exciton binding energies.

energies. By directly creating the exciton (or pair system, see below), the role of site migration is reduced. For the important case of donors, the exciton binding energies are related to the electron binding energies by Hund's rule. In turn the differences in donor binding energies between different chemical species depends on the individual size of the central cell correction to the Coulomb binding of the electron. In fact resonant excitation alone turns out to be insufficient to discriminate donor transitions, and magnetic field application must generally be used for this.

In fact resonant excitation has been employed in several situations, and before reviewing the application to donor bound exciton spectra, it will be instructive to mention two other applications. Although different acceptor chemical species can be distinguished  through bound exciton spectra, clear mapping of more subtle properties like the acceptor excited state levels have been best achieved using resonance methods. One of the best examples is the use of "selective pair" spectroscopy, in which specific pair separations from within the possible wide distributions of donor-acceptor pairs are populated by the laser probe. For a particular pair, it is possible to resonantly create an excited electron and ground state hole (or vice versa). The excited pair rapidly thermalise to the total ground state configuration, and recombines radiatively to give a sharp line characteristic of a single pair. The luminescence line is shifted from the pump line by the separation between the ground and excited states, allowing one to perform excited state spectroscopy. This elegant method has been described by several authors (12), (13).

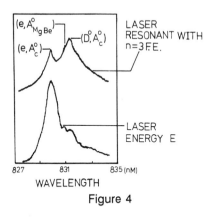

WAVELENGTH

Figure 4

A recent extension to the use of resonant excitation has been to show that the resonant creation of free excitons can dramatically change the observed PL spectrum (14). This shows that the capture cross section of the exciton may vary significantly between different impurity sites. Fig. 4 (from ref 14) shows that the PL spectrum from p-type GaAs shows normal p-type characteristics, ie. dominated by the carbon free to bound transition, when excited non-resonantly by above gap light. When the excitation was tuned to the n=3 excited state of the free exciton, the spectrum is actually dominated by the donor-acceptor pair band. It was suggested that the large spatial extent of the n=3 exciton wave function, allows efficient coupling into the donor acceptor pair complex. This interesting result broadens the notion of resonant spectroscopy to include resonance of the exciton capture processes.

Returning to the donor bound exciton problem, it has emerged in recent years (14) (15) that resonant excitation (to induce site selectivity and hence line narrowing) spectroscopy of two electron satellite spectra measured in strong magnetic fields is the most successful route to species discrimination. The concept of two electron satellites transitions is important for this task. The decay scheme is illustrated in Fig. 5. The initial state of a normal donor bound exciton transition consists of the exciton with its rather extended wave function bound to the donor; the donor is neutral with the bound electron in the 1s ground state. After decay, the donor electron remains in the 1s ground state. In the case of the satellite transition, the donor bound electron is left in an excited state after the exciton collapse. The illustrated scheme shows the final state as one of the magnetically split 2p states. Indeed a strong magnetic field is important: it compresses the electron wave function onto the impurity core, enhancing the central cell corrections or chemical shifts. This extra localisation reduces further the line widths aiding the resolution of these narrow lines, and reducing site migration.

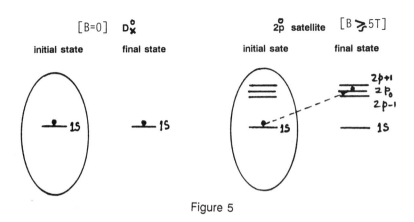

Figure 5

41

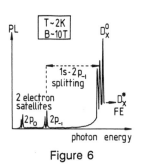

Figure 6

The splitting between the main bound exciton line and its satellite ie the 1s-2p separation is a sensitive measure of the chemical shift, representing 7/8 of the total shift. This results from the weakness of the 2p wave function at the nuclear core, compared to that of the 1s state. An example of such a spectrum for high quality InP is shown in Figure 6. The spectrum was obtained by resonantly exciting at the energy of the free exciton. Such spectra are somewhat complicated by the fact that the initial state may be one of several excited states, but the red shifting of the satellite transitions with respect to ground state exciton lines is clear. Of course the zero field chemical shifts must be obtained by extrapolation.

## 4 PHOTOLUMINESCENCE AND P.L.E. OF QUANTUM WELLS

In low dimensional structures, it is well known that the free exciton binding energy is increased and free exciton recombination often dominates the optical spectra. It is also well proven that in most material systems quantum wells are optically very bright; carrier collection into a quantum well can be very efficient. The well usually robs the surrounding barrier material of photo-generated carriers to the extent that it may be difficult to observe PL from the barrier material of a multi-quantum well stack, even the volume of the barrier material may be orders of magnitude greater than that of the quantum wells. Although sophisticated experiments measuring, for example, the transfer of excitons between coupled quantum wells have been developed (16), one of the key uses of PL, and which remains topical, has been characterisation of the abruptness of the interface systems. The basic

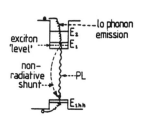

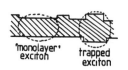

Figure 7

concept is that since the confinement energy the electron sub-bands is a sensitive function of the well thickness, then the recombination energy of an exciton reflects the average thickness of the well, or at least that of the region of the well where the recombination event occurs. Thus the overall line-width, representing the convolution of all possible transitions, must reflect the width fluctuations. In fact the real situation is more complex, (17) (18), the basic recombination scheme being shown in figure 7. The width fluctuations are best considered as potential trapping sites to which excitons migrate and subsequently recombine, provided the fluctuation is comparable in length to the "size"of the exciton. Then the PL line width represents the statistical distribution of the fluctuations. Two sorts of exciton can be imagined, one which recombines at a trapping site, contributing to a fluctuation broadened line, and one which recombines in a smooth area of the well, smooth on a scale larger than the exciton orbit. Recombination of such excitons should lead to narrow emission directly related to the local well width. If the width fluctuations are very small compared to the exciton diameter, then the averaging effect of the exciton volume leads once more to smaller line widths. Thus very rough interfaces can produce narrow spectra when the roughness is on a small scale.

The overall picture relating Pl line-width from quantum wells is given in Figure 8. The line width depends strongly on both quantum well width, and on the characteristic size of the fluctuations. In the limit of ultra smooth interfaces, very narrow wells with associated sensitivity of confinement energy may exhibit so called monolayer splitting, with the exciton band breaking up into sharp peaks, which are attributed to recombination in large smooth islands of the quantum well which vary in thickness by monolayer steps. Such spectra have been reported, although some conflicting evidence has been reported recently, (19) (20).

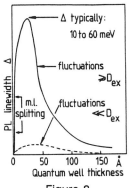

Figure 8

For the case of GaAs quantum wells, alloy broadening of the PL line width is often not an issue, because the wave-functions are largely confined to the binary material. However since the barriers are not infinite, the exciton wave functions do penetrate into the barrier, and the penetration is most severe for narrow wells. For other materials systems in which the barriers are binary and the wells ternary, the effect of alloy broadening may be more apparent (21), and it has been recorded by several groups that for (say) InGaAs/GaAs quantum wells, the narrower wells produce the narrowest line-widths. This has been interpreted as a lowering of alloy broadening as an increasing fraction of the wave function samples the binary semiconductor.

It is shown in Figure 7 that carriers or excitons collected by a quantum well thermalise by LO phonon emission before recombining at (the exciton) ground state. This means that the PL emission really only explores the ground state, and information on the higher lying sub bands. This is unfortunate because the higher states are more sensitive to phenomena like width fluctuations and so a knowledge of the position of these states is, in principle, extremely useful for characterising quantum wells. This failing has been in large part overcome by the use of photoluminescence excitation spectroscopy (PLE). The method is illustrated in Figure (9). PLE is another example of resonant excitation. In its conventional form, a tunable laser is used to sequentially excite through the absorbtion bands of the quantum well. The luminescence is observed from the ground state transition, as before. This is unavoidable. However when the laser energy is resonant with an exciton level, the usual resonance feature observed in absorbtion is detected in the PL emission. In fact that is fundamentally what the measurement gives: the absorbtion spectrum of the quantum well, replicated in the strength of the PL.

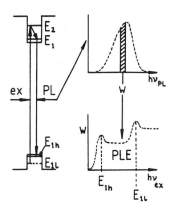

Figure 9

The PLE technique allows one to measure all of the allowed inter sub-band absorption transitions, and therefore to make a much more complete characterisation of the low dimensional system. The power of the method relies to some extent on the fact that the method effectively removes the problem of mixing in absorption in the mass of substrate material underlying the very thin epitaxy; only the epitaxial layer is measured, and in practice only the low dimensional structure contributes significantly to the PLE signal. As shown in the inset to figure 9, it is normal to select a window in the PL spectrum to constitute the PLE signal, often this is selected to be on the low energy side of the spectral peak.

A large amount of analysis has been made of quantum wells using PLE ref (22) and refs therein illustrate the detail. However there is some recent evidence to suggest that the relationship between the PL and PLE spectra may have some subtleties which are not yet fully understood. It has been noted that the PLE spectra of extremely high quality GaAs quantum wells may exhibit PLE spectra which indicate the presence of holes in the wells; information which would not of been gleaned from the excellent PL spectra (23).

In some recent work on the PLE of lattice matched InGaAs-InP quantum wells (24), there is further evidence of that PLE and PL may give apparently conflicting information on the optical and electronic properties low dimensional systems. The inter sub-band transitions of InGaAs-InP quantum wells are too low in energy to be conveniently accessed by tunable laser sources. Using a monochromator has been the traditional route for tuning the wavelength of the excitation source. However the unavoidably low brightness of such a source leads to rather low PL signal strengths, and in order to overcome this problem, a very large fraction of the PL band is included in the PLE detection window. A more sensitive method for obtaining these spectra is to use a Fourier Transform spectrometer as the excitation source, (25).

Using this method it was possible to use narrow detection windows, and to move the window through the PL band, measuring PLE spectra for each window. In figure 10 two spectra are shown. One spectrum shows strong absorption transitions for both the n=1 and n=2 heavy hole states. This is what might be expected of such quantum wells, and this spectrum was taken with the detection window on the low energy side of the PL peak. The other spectrum was measured with the detection window on the high energy side of the PL emission band. Now a quenching of the n=1 transition is observed. The bleaching effect seems to increase rather smoothly as the detection window is moved through into the high energy tail, of the PL. The implication of these data is that the overall luminescence regions in the plane of the quantum well which are of different quality. The bleaching of the n=1 transition is probably associated with regions of the well which are significantly electron occupied. The fact that different regions of the wells different characteristic absorption spectra, simply means that excitons are not free to move into all parts of the well; if the excitons were free, the n=1 absorbtion process in the high quality regions would feed luminescence in the carrier rich regions. The low diffusivity of excitons at low temperature in such mixed crystal quantum wells is thought to arise from composition fluctuation induced exciton trapping.

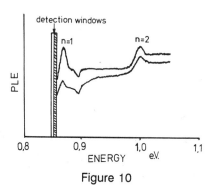

Figure 10

# REFERENCES

(1)      K. Leo and W.W. Ruhle.  Solid State Comm. $\underline{62}$, 659 (1987)

(2)      M. Tajima.  Appl. Phys. Lett. $\underline{32}$, 719 (1988)

(3)      Tamor and Wolf.  Phys. Rev. B. vol.38, 8, p.5628 (1988)

(4)      Hilmer, Mayer, Forchel, Lochner and Bauser.  Appl. Phys. Lett. **49** (15) p.948 (1986)

(5)      Hilmer et al.  Appl. Phys. Lett. **53** (20) p.1937 (1988)

(6)      Smith, Preston, Wolfe, Wake, Klem, Henderson and Morkoc.  Phys. Rev. B. vol.39 (3), p.1862 (1986)

(7)      R.J. Nelson and R.G. Sobers.  Appl. Phys. Lett. $\underline{32}$ (11), p.761 (1978)

(8)      A. Many, Y. Goldstein, N.R. Grover. "Semiconductor Surfaces" North Holland, (1971)

(9)      P. Dawson and K. Woodbridge.  Appl. Phys. Lett. $\underline{45}$ (11), p.1227 (1984)

(10)     Dean.  Prog. Cryst. Growth Charact. **5**, p.89 (1982)

(11)     U. Kaufman and J. Schneider.  Solid State Commun. $\underline{20}$, p.143 (1976)

(12)     Tews, Venghausen and Dean.  Phys. Rev. B. **19** p.5178 (1979)

(13)     Cavins, Yeo and Hengehold.  J. Appl. Phys. **64** (12) p.661 (1988)

(14)     Reynolds, Colter, Litton and Smith.  J. Appl. Phys. **55** p.1610 (1984)

(15)     S.S. Bose, B. Lee, M.H. Kin and G.E. Stillman.  Appl. Phys. Lett. $\underline{51}$ (12) p.937 (1987)

(16)     Clerot, Deveaud, Chomette Regreney and Sermage.  Phys. Rev. B. **41** (9), p.5756 (1990)

(17)     Skolnick et al.  Semicond. Sci. and Technol. $\underline{1}$, p.29 (1986)

(18)     J. Singh and K.D. Bajaj.  J. Appl. Phys. $\underline{57}$ (12) p.5433 (1985)

(19)     Bastard et al.  Phys. Rev. B. $\underline{29}$ (12) p.7402 (1984)

(20)     A. Ourmazd et al.  Mat. Sci. Forum. $\underline{38}$ to $\underline{41}$ p.689 (1989)

(21)     J. Singh and K. Bajaj.  Appl. Phys. Lett. $\underline{44}$ p.1075 (1984)

(22)     P. Dawson et al.  IOP Conf. Ser. $\underline{74}$ p.391 (1985)

(23)     Miller and Bhat.  J. Appl. Phys. **64**, p.3654 (1988)

(24)     Hamilton et al.  To be published.

(25)     Hamilton et al.  Semicond. Scie. and Technol. $\underline{3}$ p.1667 (1988)

FABRICATION OF ARTIFICIALLY LAYERED III-V SEMICONDUCTORS BY BEAM EPITAXY

AND ASPECTS OF ADDITIONAL LATERAL PATTERNING

Klaus  Ploog

Max-Planck-Institut für Festkörperforschung
D-7000 Stuttgart-80, Germany

1. INTRODUCTION

The improvement of epitaxial growth techniques has reached a status where monolayer dimensions in artificially layered semiconductor crystals are being routinely controlled to form a new class of materials with accurately tailored electrical and optical properties / 1 /. The unique capabilities of the advanced epitaxial growth techniques in terms of spatially resolved materials synthesis has stimulated the inspiration of device engineers to design a whole new generation of electronic and photonic devices based on the concept of band-gap engineering / 2 /. This concept, also called wavefunction or density-of-states engineering, respectively, relies on the arbitrary modulation of the band-edge potential in semiconductors through the abrupt change of composition (e.g. GaAs/AlAs, $Ga_xIn_{1-x}As/InP$, GaSb/InAs, Si/Ge, etc.) or of dopant species normal to the growth surface / 3 /. Applying additional lateral confinement to these artificially layered materials via lithographically defined submicron patterns has produced a new class of exotic semiconductor structures in which the quantum-mechanical properties of the electron (hole) can be fully exploited. The microscopic structuring or engineering of semiconducting solids to within atomic dimensions is thus achieved by the incorporation of interfaces (consisting of abrupt homo- or heterojunctions) into a crystal in well-defined geometrical and spatial arrangements. The electrical and optical properties of these low-dimensional semiconductor structures are then defined locally, and phenomena related to extremely small dimensions ("quantum size effects") become more important than the actual chemical properties of the materials involved.

In this article we first discuss the fundamentals of some advanced beam epitaxy techniques for the growth of III-V compound semiconductors, including conventional molecular beam epitaxy (MBE) and the various subcategories of gas-source molecular beam  epitaxy (GS MBE). We then present a few selected examples for the synthesis of artificially layered III-V semiconductors, in which we change the band structure over distances as small as a few layers of atoms in the direction of growth. We demonstrate that in nanometer-scale man-made heterostructures, such as quantum wells, tunneling barriers, and superlattices, we are able to exploit the quantum-mechanical properties of the electron. Over distances on the order of the size of atoms, which are characteristic for the length scale of artificially layered semiconductor structures of today, the electrons behave like waves in the sense that they show interference effects. Next we present a few examples for the new class of exotic semiconductor microstructures, including quantum wires and quantum dots, which are produced by additional lateral confinement using lithographically defined or focussed-ion-beam-written submicron patterns imposed on quantum-well structures. Finally, we show in which way the nearly arbitrary spatial control of the composition and doping in artificially structured semiconductors can improve the performance of conventional electronic and photonic devices. In addition, the incorporation of quantum wells, tunneling barriers, and superlattices into the active region of devices and their further lateral quantum confinement will result in the evolution of novel nanoscale devices with ultrafast switching and response time.

## 2. FUNDAMENTALS OF SOLID-SOURCE (CONVENTIONAL) AND OF GAS-SOURCE MOLECULAR BEAM EPITAXY

The fabrication of custom-designed semiconductor microstructures, where the desired potential differences are defined locally by the accurate positioning of heterojunctions on a nanometer scale, requires advanced epitaxial growth techniques, such as molecular beam epitaxy / 4 / or metalorganic vapour phase epitaxy (MO VPE) / 5 /. The particular merits of these crystal growth techniques are that ultrathin layers can be grown with precise control over thickness, composition, and doping level. Molecular beam epitaxy using elemental sources is in principle a rather elementary synthetic process, because only surface phenomena are involved in the crystal growth, no foreign atoms are present at the gas-solid interface, and by-products are not formed. Our understanding of the basic growth mechanisms involved in elemental source MBE (ES MBE) is very advanced / 6 /, and techniques to monitor and control atomic layer deposition are readily available [reflection high energy electron diffraction (RHEED) and reflectance difference spectroscopy (RDS)]. On the contrary, the basic processes occurring in MO VPE, that involve complex combinations of gas phase and surface phenomena, are at present not well understood, and monitoring techniques are only currently being developed / 5 /. In addition, since compounds are used as precursors in MO VPE, a great effort has to be expended in order to achieve the required levels of high purity and safety convenience of the growth process.

The use of any gaseous sources, whether the group III or group V in the form of alkyls and/or hydrides, in conventional MBE systems has introduced a further dimension to molecular beam epitaxy / 6, 7 /. The potential versatility of source material can now be combined with the extremely sharp interfaces and the availability of monitoring techniques associated with beam epitaxy. Depending on the various combinations of sources, the subcategories of beam epitaxy can be classified as illustrated in Fig. 1. Although different classification schemes and differences in the terminology exist, we use the convention that the term gas-source MBE (GS MBE) denotes the use of any gaseous sources whether group III and/or group V to generate the beams for epitaxy. The distinction between beam epitaxy and vapour phase epitaxy (VPE) is then made primarily about the pressure regimes involved in each method. Reasonable criteria for beam epitaxy are that the pressure be low enough that the transport of atoms and molecules is by molecular flow and that the minimum free path to be no less than the source-to-substrate distance. The upper pressure limit for molecular flow in a typical MBE system is found to be about $10^{-3}$ Torr, in agreement with simple criteria provided earlier by Dushman / 9 / for determining the upper pressure limit for molecular flow and the lower pressure limit for viscous flow of air. Although the precise pressure limits are not sharply defined, under molecular-flow conditions there is clearly no boundary layer at the substrate-gas interface. The loss of this boundary layer may imply,

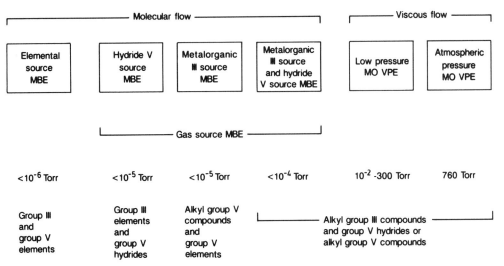

Fig. 1 Classification of vapour phase epitaxy and of the various subcategories of beam epitaxy according to the pressure regimes involved in each method.

however, that useful boundary layer reactions which limit carbon incorporation from the metal alkyls may also be lost. Although there are significant differences in the complexity of gas handling systems for toxic and hazardous chemicals and in the growth mechanisms between the various beam techniques, several GS MBE subcategories now compare favourably with ES MBE particularly for those III-V semiconductors containing phosphorus / 10 /.

## 2.1 Elemental-source (conventional) molecular beam epitaxy

The MBE technique allows lattice-plane by lattice-plane deposition of custom-designed microstructures in a two-dimensional (2D) growth process / 11 /. Crystalline materials in alternating layers of arbitrary composition and only a few atomic layers thick can thus be fabricated in a reproducible manner / 12 /. Of particular importance is that the mono-crystalline pattern of the lattice unit in successive layers continues without disruption or distortion across the interfaces between the layers. The basic process for ES MBE growth of III-V semiconductors consists of a co-evaporation of the constituent elements (Al, Ga, In, P, As, Sb) of the epitaxial layer and of dopants (mainly Be for p-type and Si for n-type doping) onto a heated crystalline substrate where they react chemically under ultra-high vacuum (UHV) conditions. The composition of the layer and its doping level depend mainly on the relative arrival rates of the constituent elements which in turn depend on the evaporation rates of the respective sources. Accurately adjusted temperatures (to within $+0.1^{\circ}$C at $1000^{\circ}$C) have thus a direct and controllable effect upon the growth process.

The group III elements are always supplied as monomers by evaporation from the respective liquid element, and they have a unity sticking coefficient over most of the substrate temperature range used for film growth (e.g. 500 – 630 $^{\circ}$C for GaAs) / 13 /. The group V elements, on the other hand, are supplied as tetramers ($P_4$, $As_4$, $Sb_4$) by sublimation from the respective solid element or as dimers ($P_2$, $As_2$, $Sb_2$) by dissociating the tetrameric molecules in a cracker of a two-zone furnace. The film growth rate is typically 0.5 - 1.0 μm/h. It is chosen low enough that dissociation and migration of the impinging species on the growing surface to the appropriate lattice sites is ensured without incorporating crystalline defects. Simple mechanical shutters in front of the evaporation sources are used to interrupt the beam fluxes in order to start and stop deposition and doping. Due to the slow growth rate of about one lattice plane per second, changes in composition and doping can thus be abrupt on an atomic scale. This independent and accurate control of the individual beam sources allows the precise fabrication of artificially layered semiconductor structures on an atomic scale, as revealed for example by transmission electron microscopy.

The stoichiometry of most III-V semiconductors during MBE growth is self-regulating as long as excess group V molecules are impinging on the growing surface. The excess group V species do not stick on the heated substrate surface, and the growth rate of the films is essentially determined by the arrival rates of the group III elements / 13 /. A good control of ternary III-III-V alloys can thus be achieved by supplying excess group V species and adjusting the flux densities of the impinging group III beams, as long as the substrate temperature is kept below the congruent evaporation limit of the less stable of the constituent binary III-V compounds (i.e. GaAs in the case of $Al_xGa_{1-x}As$). Hence, at higher growth temperature preferential desorption of the more volatile group III element (i.e. Ga from $Al_xGa_{1-x}As$) occurs so that the final film composition is not only determined by the added flux ratios but also by the differences in the desorption rates. To a first approximation we can estimate the loss rate of the group III elements from their vapour pressure data. The surface of alloys grown at high temperatures will thus be enriched in the less volatile group III elements. As a consequence, we observe a significant loss of In in $Ga_xIn_{1-x}As$ layers grown above 550 $^\circ$C and a loss of Ga in $Al_xGa_{1-x}As$ layers grown above 650 $^\circ$C. The growth of ternary III-V-V alloys, like $GaP_yAs_{1-y}$, by MBE is more complicated, as even at moderate substrate temperatures the relative amounts of the group V elements incorporated into the growing layer are not simply proportional to their relative arrival rates. The factors controlling this incorporation behaviour are at present not well understood / 13 /, also not for the other beam epitaxy methods depicted in Fig. 1. It is therefore difficult to obtain a reproducible composition control during ES MBE growth of ternary III-V-V alloys, especially when the tetrameric group V species are used. In ternary III-III-V alloys, on the other hand, the simplicity of the MBE process allows compositional control from x = 0 to x = 1 in $Al_xGa_{1-x}As$, $Ga_xIn_{1-x}As$ etc. with a precision of ±0.001 and doping control, both p- and n-type, from the $10^{13}$ cm$^{-3}$ to the $10^{19}$ cm$^{-3}$ range with a precision of a few percent. The accuracy is largely determined by the care with which the growth rate and doping level were precisely calibrated in test layers.

The purity of most III-V semiconductors grown by ES MBE is limited by background impurities originating from the UHV system and from the source materials. Not intentionally doped GaAs layers with residual impurity concentrations in the low $10^{13}$ cm$^{-3}$ range are now routinely achieved. The most common dopants used during MBE growth are Be for p-type and Si for n-type doping. The group II element Be behaves as an almost ideal shallow acceptor in many MBE grown III-V semiconductors / 14 /. Each incident Be atom produces one ionized impurity species, providing an acceptor level 29 meV above the valence band edge in GaAs. At doping concentrations above $3 \times 10^{19}$ cm$^{-3}$, however, the GaAs surface morphology and electronic properties degrade, and the diffusion of Be is enhanced when the samples are grown at substrate temperatures above 550 $^\circ$C / 15 /. Lowering the substrate temperature makes feasible Be acceptor concentrations above $10^{20}$ cm$^{-3}$ in GaAs with smooth surface morphology. The group IV element Si is primarily incorporated on group III sites during MBE growth under As-stabilized

conditions, yielding n-type material of fairly low compensation. The observed doping level is simply proportional to the dopant arrival rate,
provided care is taken to reduce the $H_2O$ and $CO$ background level during
growth / 16 /. The upper limit of $n = 1 \times 10^{19}$ $cm^{-3}$ for the free-electron
concentration in GaAs is given by the formation of [Si-X] complexes, where
X is assigned to Ga vacancies ($V_{Ga}$) which account for the compensation
found in very heavily doped layers / 17 /. The possibility of Si migration
during MBE growth of $Al_xGa_{1-x}As$ layers at high substrate temperatures and/
or with high donor concentrations has been the subject of controverse discussions, because of its deleterious effects on the properties of selectively doped $Al_xGa_{1-x}As$/GaAs heterostructures. There is now sufficient evidence that at high doping concentration ($> 10^{18}$ $cm^{-3}$) the Si migration in
$Al_xGa_{1-x}As$ layers results from a concentration-dependent diffusion process
which is enhanced at high substrate temperatures / 18 /. It is finally
noteworthy that the incorporation of Si atoms on either Ga or As sites
during MBE growth depends strongly on the orientation of the GaAs substrate / 19 /. In GaAs deposited on (111)A, (211)A and (311)A orientations
the Si atoms predominantly occupy As sites and act as acceptors, whereas
they occupy Ga sites and act as donors on (001), (111)B, (211)B, (311)B,
(511)A, (511)B and higher-index orientations.

The advanced MBE systems mostly consist of three basic UHV building
blocks, i.e. the growth chamber, the sample preparation chamber, and the
load-lock chamber, which are separately pumped and interconnected via large-
diameter channels and isolation valves / 20 /. High-quality layered semiconductor structures require a background vacuum in the low $10^{-11}$ Torr
range to avoid incorporation of background impurities into the growing
layers. Therefore, extensive $LN_2$ cryoshrouds are used around the substrate to achieve locally much lower background pressures of condensible species. The starting materials for the growth of III-V semiconductors are
evaporated in resistively heated effusion cells made of pyrolytic BN which
operate at temperatures up to 1400 °C. Most of the functions important for
the MBE growth process are controlled by a computer.

Molecular beam epitaxy of III-V semiconductors is mostly performed
on (001) oriented substrate slices about 300 - 500 μm thick. The preparation of the growth face of the substrate from the polishing stage to the
in-situ cleaning stage in the UHV system is of crucial importance for epitaxial growth of ultrathin layers and heterostructures with high purity
and crystal perfection and with accurately controlled interfaces on an
atomic scale. The substrate surface should be free of crystallographic defects and clean on an atomic scale. Various cleaning methods have been
described for GaAs and InP / 20 /, which are the most important substrate
materials for deposition of III-V semiconductors. The first step always
involves chemical cleaning and etching, which leaves the surface covered
with some kind of a protective oxide. After insertion in the MBE system
this oxide is removed by heating under UHV conditions carried out in a
beam of the group V element.

The most important method to monitor in-situ surface crystallography
and growth kinetics during MBE is reflection high-energy electron diffraction (RHEED) operated at 10 - 50 keV in the small glancing angle reflection mode / 21 /. The diffraction pattern on the fluorescent screen, mostly taken in the [100] and [110] azimuths of (001) oriented substrates,
contains information from the topmost layers of the deposited material ,
and it can thus be related to the topography and structure of the growing
surface. The diffraction spots are elongated to characteristic streaks
normal to the shadow edge. Additional features in the RHEED pattern at
fractional intervals between the bulk diffraction streaks manifest the
existence of specific surface reconstructions, which are correlated to the
surface stoichiometry and thus directly to the MBE growth conditions
(substrate temperature, molecular beam flux ratio etc.) / 11 /.

Another characteristic feature of the RHEED pattern is the existence of pronounced periodic intensity oscillations of the specularly reflected and of several diffracted beams during MBE growth / 22, 23 /. The period of these oscillations corresponds exactly to the time required to deposit a lattice plane of GaAs (or AlAs, $Al_xGa_{1-x}As$, etc.) on the (001) surface. This peculiarity is explained by the assumption that the amplitude of the intensity oscillation reaches its maximum when the lattice plane is just completed (maximum reflectivity). The formation of the following lattice plane starts with statistically distributed 2D islands having the height of one GaAs lattice plane [0.28 nm for the (001) plane]. The intensity of the reflected (or diffracted) electron beam decreases with increasing size of the islands. The minimum reflectivity occurs at half-layer coverage ($\Theta = 0.5$). When the coverage is further increased the islands coalesce more and more, and the reflectivity reaches a maximum again at $\Theta = 1$. The observed intensity oscillations in the RHEED pattern provide direct evidence that MBE growth occurs predominantly in a 2D layer-by-layer growth mode. This method is now widely used to calibrate and to monitor absolute growth rates in real time with monolayer resolution. However, it is important to note that the interpretation of RHEED intensity data may be erroneous because in general both elastically and diffusely scattered electrons contribute to the recorded intensity. Diffraction conditions where elastic scattering dominates the intensity of the specularly reflected beam are most easily obtained under off-azimuth conditions. The other important aspect is the damping of the RHEED intensity oscillations observed during growth and the characteristic recovery once growth has been interrupted The damping has been ascribed / 24 / to an increase of the surface step density, i.e. an increasing number of individual surface domains probed by the electron beam are no longer in phase. The recovery of the RHEED intensity following interruption of growth has been identified with an expansion of the mean terrace width of the surface and hence a reduction of the surface step density, i.e. a few large domains on the growing surface were formed.

Recently, an in-situ optical technique called reflectance difference spectroscopy (RDS) has been introduced / 25 / to monitor the surface chemistry during MBE growth of GaAs and $Al_xGa_{1-x}As$ layers and heterostructures. This technique is primarily sensitive to the anisotropy of the growing surface induced by Ga-Ga dimers and hence to the surface stoichiometry. In combination with RHEED a careful analysis of RDS data provides detailed insights into the MBE growth mechanisms of III-V compounds.

## 2.2  Gas-source molecular beam epitaxy

In the various subcategories of gas-source molecular beam epitaxy (GS MBE) the elemental sources are replaced by volatile metalorganic compounds and/or hydrides to transport the constituents to the growing gas-solid interface. In general the trimethyl- or triethyl-group III compounds (TMAl, TMGa, TMIn, TEAl, TEGa, TEIn etc.) are used in combination with the group V hydrides $PH_3$ and $AsH_3$. The replacement of the elemental sources in conventional MBE by gaseous source materials was initiated by the search for a long-lasting arsenic source and for a reproducible composition control during growth of ternary III-V-V alloys based on phosphorus and arsenic / 26 /. Thermal cracking of the hydrides $PH_3$ and $AsH_3$ at temperatures around 900 °C produces the dimers $P_2$ and $As_2$ as well as hydrogen. This hydride source MBE (HS MBE) / 7 /, where only the group V elements are replaced by their respective hydrides, allows epitaxial growth of $Ga_{0.47}In_{0.53}As$/InP heterostructures and superlattices of high quality and of $Ga_xIn_{1-x}P_yAs_{1-y}$/InP heterostructures with reasonable composition control.

An extension of this concept was then made by replacement of the group III elements by metalorganic compounds / 27 /. In this metalorganic MBE (MO MBE) and chemical beam epitaxy (CBE) / 8 / the metalorganic flows mixed with hydrogen are in some cases combined outside the UHV growth chamber to form a single beam impinging onto the heated substrate for a good compositional uniformity across the substrate area. On the heated substrate surface thermal pyrolysis of the metalorganic compounds takes place and in an excess group V beam the III-V semiconductor is formed. Also these techniques allow reproducible growth of P containing III-V semiconductor heterostructures of high quality.

In UHV systems used for GS MBE the effusion cells are replaced by metalorganic entry tubes and/or by the group V gas source cracker. To provide for adequate pumping of the large amounts of hydrogen, a throughput turbomolecular or diffusion pump of sufficient pumping speed has to be attached to the UHV system. One of the most important components of GS MBE systems is the gas handling system that provides means of regulating the flow rates of the starting compounds into the growth chamber. Precision mass flow controllers with linear response are used to control directly the flow of the pure gaseous group V hydrides into the thermal cracker. The metalorganics are transported by hydrogen carrier gas of fixed pressure bubbling through the (liquid) metalorganics kept at accurately controlled temperatures. Mass flow control is then of the combined gas stream. For doping the elements Be and Si evaporated in conventional effusion cells are used also in gas-source molecular beam epitaxy / 7, 8 /.

It is obvious that there are significant differences in growth mechanisms between the various subcategories of MBE, due to the use of metalorganic species / 28 /. In solid-source and hydride-source MBE there is no interaction in the beams, and the growth rate depends only on the arrival rate of the (elemental) group III species / 13 /. Since these have unity sticking coefficients up to a certain critical temperature, the growth rate is relatively independent of substrate temperature. In MO MBE and CBE, on the other hand, the substrate temperature adopts two functions. Firstly, it has to decompose the metalorganics and secondly, to impart energy to ensure sufficient adatom mobility for epitaxial growth. The growth rate now depends on substrate temperature. At low temperatures, the growth rate is limited by the efficiency of the pyrolysis of the metalorganics. Then a plateau range exists where this pyrolysis occurs at a constant rate. Unlike in ES MBE, the mobile surface species is a metal alkyl with at least one alkyl group removed. Recent studies of the surface chemistry confirm that the cleavage of the metal-alkyl bonds occurs sequentially. This adsorbed metal-alkyl species may have an even greater surface mobility at a given substrate temperature than its elemental counterpart. At higher substrate temperatures, finally, a similar re-evaporation process occurs to that of the elemental species. As a consequence, the growth rate and also the residual (carbon) impurity concentration depend on the substrate temperature and on the group-V/group-III flux ratio in a very complicated manner / 7, 8, 28 /.

A distinct example for the differences in growth mechanisms between ES MBE and GS MBE and for our poor understanding of the latter is the epitaxy of $Ga_x In_{1-x} As$ lattice matched to InP substrate / 28 /. The difference is exemplified by the variation of In concentration with substrate temperature. In ES MBE the In concentration remains constant until loss of the least stable binary compound (i.e. InAs) occurs at higher temperatures. A similar behaviour, but over a wider temperature range, is obtained for MO VPE. However, the composition vs. temperature profile changes noticeably with the use of metal alkyls in GS MBE. At low temperature, the pyrolysis of the Ga alkyl is less efficient as compared to the In alkyl, resulting

in an increased In content in the $Ga_xIn_{1-x}As$ alloy. When the substrate
temperature is increased, the pyrolysis rates stabilize and lattice matching of the $Ga_xIn_{1-x}As$ with x = 0.47 is achieved, but only over a very
narrow temperature range. When the temperature is further increased, the
In concentration in the alloy increases rapidly. This trend is opposite to
that observed in ES MBE and is not yet understood. Obviously, this change
in alloy composition is caused by an anomalous enhancement of the Ga alkyl
desorption in the presence of an incident In alkyl flux. This problem of
competing surface reactions highlights our current poor understanding of
surface chemistry involved in MO MBE and in CBE.

When the decomposition of the metal alkyls on the growing surface is
not completed, carbon may be incorporated into the growing layer / 8, 27 /.
This tendency for the grown material to be heavily p-type is more pronounced with TMG as starting material. In the case of TEG the so-called ß-elimination process facilitates the breaking of the Ga-carbon bond so that
the background doping level in GaAs can be reduced to the low $10^{14}$ $cm^{-3}$
range / 8 /. In general, starting metalorganics having a weaker bonding
of the metal alkyl and allowing for ß-elimination are advantageous for MO
MBE. However, whether the residual carbon incorporates with the group III
radical or from chemisorbed organic radicals is not yet clarified. Also
in the various subcategories of gas-source molecular beam epitaxy RHEED
and RDS are now more frequently used to monitor in-situ the surface chemistry and growth mechanisms underlying the epitaxial growth / 25, 28 /.
These investigations will hopefully improve our present poor knowledge
in this field.

The various subcategories of GS MBE close the gap between the techniques of conventional ES MBE and low-pressure MO VPE, as indicated in Fig.
1. These more recent developments have advantages for the growth of III-V
semiconductors containing phosphorus. Although a number of impressive results have been achieved, a detailed investigation of the incorporation
behaviour of phosphorus and arsenic in ternary III-V-V compounds during
HS MBE and CBE is still lacking. In addition, a thorough comparison of the
properties of Al containing heterostructures and superlattices grown either
by ES MBE and HS MBE or by MO MBE and CBE would reveal the actual state
of the art of each of these techniques. Finally, the real challenge for
both GS MBE and MO VPE is the replacement of the extremely toxic $AsH_3$ by
suitable safer arsenic compounds of high purity.

## 2.3  Modulated beam techniques

In the last three years, the original methods of beam epitaxy have
encountered important new developments to expand their application particularly to low-temperature and hetero-epitaxial (lattice-mismatched) growth.
While in the conventional techniques the molecular or metalorganic beams
impinge continuously onto the substrate surface for the growth of a homogeneous layer (e.g. GaAs), alternating or modulated beams synchronized
with the layer-by-layer growth mode by properly actuating the shutters are
now used in migration-enhanced epitaxy (MEE) / 29 / and in atomic-layer
MBE (AL MBE) / 30 /, which will be discussed here. To a certain extent
these modulated beam techniques artificially induce an atomic layer-by-layer
growth sequence on the (001) surface. As shown in Table 1, their advantages
are the superior smooth morphology of surfaces and interfaces especially
for lattice-mismatched (strained) systems, the suppression of oval-defect
formation, and the substantial lowering of the favorable growth temperature
for high-quality material. In MEE both the group III and the group V molecular beams are alternating with a periodicity corresponding to the atomic
layer-by-layer growth sequence. The group III dose per cycle has to be
carefully adjusted to that required for completion of one atomic layer by

using RHEED intensity oscillation profiles as reference. The disadvantages of MEE are the low growth rate and the high mechanical load for the shutters. In the AL MBE method, on the other hand, the group III beam is impinging continuously onto the substrate while periodic short pulses of group V species are supplied with a repetition rate synchronized with the monolayer growth rate. Also in this case the atomic layer-by-layer growth sequence is accomplished, provided that (i) the group V repetition rate coincides accurately with the growth rate in monolayers per second, (ii) the group V pulses are short, and (iii) their intensity is adequate to saturate the surface at each cycle. The advantages of AL MBE as compared to MEE are thus a higher growth rate (comparable to that of conventional MBE) and less mechanical load on the shutters. The growth mechanism of both MEE and AL MBE is based on an instantaneous group III enrichment of the (001) surface during each cycle, which can be monitored by a dramatic change of the surface reconstruction via the RHEED pattern. The growing surface thus alternates between a group III completed and a group V saturated state. According to Horikoshi et al. / 29 / the growth mechanism of MEE involves the enhancement of the surface migration of the group III species when supplied separately. Briones et al. / 30 /, however, assume a mechanism of enhanced nucleation and forced 2D growth for AL MBE which is

TABLE 1    Comparison of the characteristics of conventional MBE with MEE and AL MBE

|  | CONVENTIONAL MOLECULAR BEAM EPITAXY (MBE) | MIGRATION ENHANCED EPITAXY (MEE) | ATOMIC LAYER MOLECULAR BEAM EPITAXY (ALMBE) |
|---|---|---|---|
| Characteristic features | Continuous flux of group III and group V elements; Substrate temperature 500°C for high-quality GaAs | Periodic modulation of both the group III and group V element flux synchronously with the atomic layer-by-layer sequence (RHEED oscillations); Growth temperature lowered by 200°C as compared to MBE | Continuous flux of group III element(s), but short pulses (modulation) of the group V flux synchronously with RHEED intensity oscillations; Growth temperature lowered by 200°C as compared to MBE |
| Advantages | Reasonable growth rates; Excellent material quality for lattice-matched systems; Reasonably abrupt interfaces | Excellent surface morphology (no "oval" defects); Low growth temperature; Abrupt interfaces in ultra-thin-layer superlattices; 2D growth in lattice-mismatched systems; Controlled incorporation of two group V elements | Excellent surface morphology (no "oval" defects); Low growth temperature; Reasonable growth rates; Abrupt interfaces in ultra-thin-layer superlattices; 2D growth in lattice-mismatched systems; Controlled incorporation of two group V elements |
| Problems | Formation of "oval" defects; Interfaces in ultrathin-layer superlattices; Strained-layer hetero-structures | Low growth rate; Accurate control of group III flux required to deposit exactly one monolayer; High mechanical load for shutters | Mechanical design of group V element shutters; Substrate rotation vs. RHEED intensity oscillations |
| Applications | Standard heterostructures; Production of material for device fabrication | Low-temperature growth on processed substrates; Controlled 2D growth of lattice-mismatched materials (e.g. GaAs-on-Si); Ultrathin-layer super-lattices with very abrupt interfaces; Layered structures composed of two different group V elements (e.g. P and As) | Low-temperature growth on processed substrates; Controlled 2D growth of lattice-mismatched materials (e.g. GaAs-on-Si); Ultrathin-layer superlattices with very abrupt interfaces; Layered structures composed of two different group V elements (e.g. P and As) |

prompted by the synchronous separation of the surface from equilibrium. For both techniques valid is probably that the periodic changes of surface stoichiometry induce a 2D growth mechanism predominantly by enhanced layer nucleation.

## 3. CHARACTERIZATION OF CUSTOM-DESIGNED SEMICONDUCTOR MICROSTRUCTURES

The characterization of semiconductors which are microscopically structured down to atomic dimensions requires analytical techniques having a high spatial resolution and a high sensitivity and accuracy. The necessity of a detailed characterization stems directly from the history of semiconductor superlattices / 31 /. First, a sound theory made it possible to predict the intriguing properties of these artificially layered semiconductors. Then, sophisticated measurement techniques were used to assess the degree to which the predictions have been fulfilled / 2, 3, 4, 11 /. The interface perfection experimentally attainable by beam epitaxy is now so high that conventional methods of depth profiling analysis, which involve sputtering to section the material combined with some means of composition determination (such as Auger electron spectroscopy or secondary ion mass spectroscopy) are no longer adequate for the resolution required for atomically sharp interfaces. Hence, our ability to grow complex multilayer structures has led to new challenges for the characterization of materials, and experimental methods and their theoretical framework are developed to meet this challenge. Improved and new techniques to characterize materials have in turn a strong impact on the evolution of methods for spatially resolved materials synthesis.

In general several different characterization methods are now used routinely to assess the specific structural and electronic properties of microscopically structured semiconductors: (i) transmission electron microscopy (TEM), (ii) double-crystal X-ray diffraction, (iii) emission and absorption spectroscopy, (iv) Hall effect and current-voltage or capacitance-voltage measurements, and (v) Raman scattering. In many cases, a combination of two or more of these methods is necessary for a clear assessment of the specific structural and electronic properties / 32 /.

## 4. EXAMPLES OF ARTIFICIALLY LAYERED III-V SEMICONDUCTORS

### 4.1 Scaling of the concept of microscopical structuring of solids to its physical limit

The concept of microscopical structuring of solids is scaled to its physical limit normal to the crystal surface in two prototype artificially layered semiconductor structures, which we have successfully fabricated by ES MBE / 33, 34 /, i.e. $(GaAs)_m/(AlAs)_n$ ultrathin-layer superlattices and delta- (or monolayer) doping in GaAs and $GaAs/Al_xGa_{1-x}As$ structures. In both the $(GaAs)_1/(AlAs)_1$ monolayer superlattice and in the delta-doped $GaAs/Al_xGa_{1-x}As$ structures the characteristic material length normal to the surface has reached a spatial extent of less than the lattice constant. Due to the indirect-gap nature of the constituent AlAs layers, the $(GaAs)_m$ /(AlAs)$_n$ ultrathin-layer superlattices exhibit novel optical properties for certain (m, n)-values / 33, 35 /. The minority-carrier lifetimes can be tailored in the range of several hundred picoseconds to several microseconds simply by the appropriate design of the superlattice configuration, i.e. the (m, n)-values, during epitaxial growth. This unique feature opens up a new field for application of these structures in lasers and in nonlinear photonic and optoelectronic devices. Measurements of the luminescence intensity under hydrostatic presssure / 36 / have shown that the mixing between $\Gamma$-like and X-like states in these ultrathin-layer superlattices is rather small. Hence, the ratio of the transition probability of electrons

56

in the $\Gamma$- and X-like state to heavy-hole states ($P^{\Gamma}/P^{X}$) increases only slightly from $1.4 \times 10^{-4}$ for the (m, n) = 17 configuration (type I superlattice) to $4.6 \times 10^{-3}$ for the (m, n) = 6 configuration (type II superlattice), in good agreement with theoretical calculations / 37 /. In properly designed type-II (GaAs)$_m$/(AlAs)$_n$ superlattices the real-space transfer times associated with the $\Gamma$-X intervalley scattering of electrons has directly been measured by femtosecond optical pump and probe spectroscopy / 38 /. In contrast to bulk material the scattering rate for this process, which connects electron states in different slabs of the layered material, is determined by the spatial overlap of the electron states in the different satellite minima.

The incorporation of a narrow buried doping channel in delta-doped GaAs layers and in GaAl/Al$_x$Ga$_{1-x}$As heterostructures leads to a major improvement of the electrical properties / 34 /. Based on this concept, nonalloyed ohmic contacts to GaAs, field-effect transistors (FETs) with very high transconductance and excellent driving capabilities, and unpinned GaAs surfaces for MOS devices have been obtained / 39 /. Also the record electron mobility in selectively doped GaAs/Al$_x$Ga$_{1-x}$As heterostructures as high as $10^7$ cm$^2$/Vs has been attained with the concept of Si-delta-doping / 40 /. An extension of the delta-doping concept is manifested by the GaAs sawtooth doping superlattices, which consist of periodic arrays of alternating Be- and Si-delta-doping sheets interspersed by undoped GaAs regions / 41 /. These sawtooth doping superlattices exhibit very interesting optical and vertical-transport properties.

## 4.2 Resonant tunneling and Wannier-Stark localization

The experimental observation of tunneling through potential barriers is a manifestation of the wave-like behaviour of matter. The resonant tunneling diode, first proposed by Chang et al. / 42 /, consists of an n$^+$- GaAs/Al$_x$Ga$_{1-x}$As/GaAs/Al$_x$Ga$_{1-x}$As/n$^+$-GaAs layer sequence to form a series of two Al$_x$Ga$_{1-x}$As potential barriers separated by a GaAs potential well. The barriers are thin enough that electrons can tunnel through them into and out of the quantum well. Resonant tunneling occurs when the bias voltage applied across the n$^+$-GaAs electrodes is such that the subband in the quantum well matches the energy of the Fermi level of the injecting electrode. We thus observe peaks in the electron transmission (current) as a function of incident electron energy (voltage). To some extent the double barrier tunneling diode can be described in terms of a Fabry-Perot resonator / 43 /. The distinct resonance enhancement of electron transmission can hence be understood in terms of constructive interference between multiply reflected waves. The leakage current due to inelastic tunneling processes is determined by the interface quality and by electron-phonon scattering. Scattering processes also weaken the resonance enhancement of the transmission. Multiply reflected waves no longer interfere constructively after having been reflected more than a few times, because scattering events destroy their phase coeherence / 43 /.

Most of the III-V heterostructures having potential barriers for tunneling have been grown by conventional ES MBE. The recent improvements in the MBE growth process has made possible very high peak-to-valley tunnel current ratios. In a double barrier diode structure made of a strained Ga$_x$In$_{1-x}$As well and AlAs barriers a ratio of 30:1 has been observed at room temperature / 44 /. An important phenomenon in the operation of resonant-tunneling diodes is the space-charge buildup within the well. This space charge gives rise to an electrostatic potential which shifts the resonant energy relative to the emitter Fermi level (electronic bistability). The stored charge plays also an important role in determining the response time of the resonant-tunneling diode. The response time is intrinsically limited by the rate at which this charge can build up or decay. As oscillators resonant-tunneling diodes have operated up to 420 GHz / 45 /, and bistable operation has been achieved with a rise time of 2 psec / 46 /. Owing to their two-terminal nature, simple resonant-tunneling diodes find only limited practical application. Therefore, the concept of the resonant-tunneling diode has to be combined with, e.g. the concept of the bipolar transistor. The various approaches in this field have shown very promising device performances / 47 , 48 /.

In multiple quantum well (MQW) heterostructures we can increase the bias voltage until the ground state of a well is brought into resonance with one of the excited states of the neighbouring wells. Under these conditions electrons tunnel from the ground state of a given well into the excited states of the next well, followed by intra-well energy relaxation (nonradiative) to the ground state / 49 /. This process is repeated many times to produce a characteristic cascade process through the entire MQW structure. This sequential resonant tunneling produces a series of peaks in the photocurrent-voltage characteristics. We have recently observed resonant tunneling in GaAs/AlAs MQW structures from the ground state in one well (1e) to the first three excited states (2e, 3e, 4e) in the next well up to temperatures as high as 260 K / 50 /. Our time-resolved photocurrent measurements clearly revealed resonantly reduced tunneling times (i.e. resonantly enhanced electron tunneling rates) of about 10 psec for the 1e $\rightarrow$ 2e process in GaAs/AlAs MQW structures with $L_z$ = 12 nm and $L_B$ = 2 nm / 51 /. In addition we found that the tunneling time changes by a factor of two when the AlAs barrier fluctuates by $\pm$ 1 monolayer only.

A semiconductor superlattice, i.e. a MQW structure with very thin barriers, is characterized by the existence of extended minibands for electrons and holes of widths $\Delta_e$ and $\Delta_h$ in the direction of layer sequence. The electron wavefunctions are thus delocalized throughout the entire superlattice. When we apply a high electric field of $F \sim 10^5$ V/cm along the superlattice direction, all electron and hole states become completely localized (Wannier-Stark localization) and interband electron-hole transitions between the respective levels are similar to those in isolated quantum wells. In moderate electric fields ($F \sim 2 \times 10^4$ V/cm), a partial localization of the extended superlattice states to a finite number of wells occurs, resulting in a splitting of the miniband into discrete energy levels separated by eFD (D = periodicity), i.e. the so-called Stark-ladder states / 52 /. The formation of Stark-ladder states in GaAs/Al$_x$Ga$_{1-x}$As and GaAs/AlAs superlattices has recently been observed by photoluminescence / 53 / and by photocurrent / 54 / measurements. These measurements have shown that in the optical transitions not only electrons centered in the same well as the holes (intrawell transition) are involved but also electrons of other states of the electronic Stark ladder centered in different wells (interwell transitions). These non-vertical transitions are blue- and red-shifted in units of eFD (ieFD, i = $\pm$ 1, $\pm$ 2, $\pm$ 3 ...) relative to the intrawell transition. The observation of optical transitions based on the Wannier-Stark localization provides direct evidence for the quantum coherence of the superlattice wavefunctions, which is strongly affected by the quality of the constituent heterointerfaces. In GaAs/AlAs superlattices we have recently observed several distinct optical transitions related to Wannier-Stark localization even at room temperature / 55 /, indicating a coherence length of at least five superlattice periods. These transitions produce multiple regions of negative differential photoconductivity which have been used to realize a multistable self-electro-optic effect device (SEED).

## 4.3 Strained-layer heterostructures and superlattices

Epitaxial growth of custom-designed semiconductor microstructures consisting of alternating layers of chemically different materials having nearly equal lattice constants is routine now in many laboratories around the world. However, close lattice-constant matching is not a fundamental limitation for the growth of high-quality multilayer structures. For lattice mismatch less than 7% it is now possible to grow a finite thickness of epitaxial layer free of crystal defects. In this commensurate growth, the lattice constant of the epilayer in the growth plane is strained to exactly match the lattice constant of the underlying substrate. As a result the stained epilayer undergoes a tetragonal distortion, and the lattice constant in growth direction is no longer equal to the in-plane one. The ability to fabricate such strained-layer heterostructures is very attractive, not only because of the large variety of material combinations that can be produced, but also because of the use of the built-in strain to tailor the bandgap and the transport properties of such systems.

Strained-layer superlattices (SLs) thus accommodate their lattice mismatch via coherent, elastic layer strains so that misfit dislocations are not created at the constituent superlattice interfaces / 56, 57 /. An important restriction for most strained-layer SLs is that the overall thickness of the epilayer must be less than the critical layer thickness to avoid the formation of misfit dislocations. Investigations of strained III-V semiconductor systems have been concentrated mainly on strained-layer quantum wells of $Ga_xIn_{1-x}As$ grown both on GaAs and on InP substrates and the corresponding SLs. The components GaAs and InAs have a lattice mismatch of 7.1%. In the $Ga_xIn_{1-x}As$/InP system the biaxial inplane strain can be varied between -3.8% (tensile strain) to +3.2% (compressive strain), by varying the InAs mole fraction from x = 1 (GaAs) to x = 0 (InAs). At x = 0.47 the ternary epilayer is lattice-matched to the InP substrate. The strained $Ga_xIn_{1-x}As$/GaAs single and multiple quantum well (SQW and MQW) heterostructures can be considered as the prototype systems for studying electrical and optical properties of III-V strained-layer structures. Transport measurements on modulation doped $Ga_xIn_{1-x}As$/GaAs SQW and MQW structures with x ranging from 0.15 to 0.25 have revealed new phenomena arising from the strain-induced splitting of the heavy- and light-hole states at the top of the valence band / 58 /. The optical properties of the strained $Ga_xIn_{1-x}As$/GaAs quantum well system depend on three contributions, namely composition, strain, and quantum size effect / 59 /. Concerning light holes, this system often behaves like a type II superlattice, i.e. the light holes are weakly confined to the GaAs barriers while electrons and heavy holes are confined within the strained $Ga_xIn_{1-x}As$ well / 60 /. In $Ga_xIn_{1-x}As$/InP strained-layer SLs a strain-induced transition from type I to type II is observed for x > 0.80 / 60 / where electrons are confined to the binary InP and holes are confined in the strained ternary $Ga_xIn_{1-x}As$ layers. This type I/type II transition is very different from the transition observed in $(GaAs)_m/(AlAs)_n$ ultrathin-layer SLs, as in $Ga_xIn_{1-x}As$/InP strained-layer SLs the transition remains direct in momentum space. The discussed strained MQW structures consisting of several strained quantum wells separated by unstrained barriers sometimes exceed the critical epilayer thickness and are thus structurally metastable. Their total strain energy is larger than the energy to form misfit dislocation. However, owing to careful choice of layer thicknesses and growth conditions, the nucleation of dislocations causing structural degradation is kinetically hindered.

The correlation between the observed radiative recombination process and the actual structural configuration of $Ga_xIn_{1-x}As$ strained layers are often not well understood. We have therefore performed an extensive study of the effects of strain and confinement (localization) on the exciton eigenstates in monolayer-InAs/GaAs quantum well structures, which represent an all-binary strained layer system / 61 /. The studied samples were grown by conventional elemental source molecular beam epitaxy on (100)GaAs substrates. The thickness and the strain state of the interspersed InAs planes were determined with high accuracy by double-crystal X-ray diffraction. This structural information on an atomic scale has provided insight in the fundamental growth processes under a large lattice mismatch. One monolayer InAs was found to grow coherently on the GaAs buffer, i.e. no misfit dislocations are generated at the highly strained InAs/GaAs heterointerface. The X-ray measurements allowed us to determine the crystal perfection of the samples with respect to the growth conditions of the InAs planes. In particular, a considerable deviation of the cation incorporation rate into the highly strained lattice plane as compared to bulk growth was found, indicating that the independent determination of the important structural parameters of the samples is necessary for a correct interpretation of the observed electronic properties. The nature of the radiative transitions and their relation to the actual structural configuration were determined by combining different spectroscopic techniques, including photoluminescence (PL), PL excitation (PLE) and photoreflectance (PR) / 62 /. We have observed heavy- and light-hole exciton features in the PLE spectra and the corresponding PR resonances up to room temperature. The new features observed in the PLE and PR spectra arise from novel localization effects. The high luminescence efficiency of the InAs quantum-well-related recombination indicates an extremely efficient carrier capture even in the submonolayer regime with trapping times in the picosecond range. A tight-binding model in the linear-chain approximation can account for the observed excitonic features in terms of particle localization at the two-dimensional potential discontinuity produced by the insertion of the InAs planes in the host GaAs matrix.

## 5. ADDITIONAL LATERAL PATTERNING OF QUANTUM WELL STRUCTURES

Additional lateral patterning of artificially layered III-V semiconductors leads to a new class of exotic microstructures, including quantum wires and quantum dots. The lateral confinement to quantum well structures can be applied with the techniques of electrostatic squeezing, lithographically defined deep-mesa-etching, or focussed-ion-beam writing. The condition for the occurrence of new phenomena in these low-dimensional semiconductor structures is that the lateral size of the active region can be made smaller than the coherence and elastic scattering lengths. In this regime the structures act as electron waveguides, because the "lateral" wire dimensions are of the order of the de Broglie wavelength and only a few laterally defined modes are occupied. In addition to the wave nature of the electron being fundamental to the phenomena under study, the density of states changes drastically from a parabolic curve for three-dimensional (bulk) semiconductors, to a staircase shape for quasi-two-dimensional (2D) structures, to a peaked curve for quasi-one-dimensional (1D) structures, and finally to a delta-function like behaviour for quasi-zero-dimensional (0D) structures.

The electrostatic squeezing technique uses specific geometries of Schottky barrier gates imposed on the 2D electron gas at GaAs/Al$_x$Ga$_{1-x}$As heterojunctions. The application of a reverse bias removes the carriers below the gated regions, and widths of submicron conducting channels can be reduced by making the gate voltage more negative. Using the split-gate geometry a range of channel widths can be explored in a single sample, and a device of length less than both the elastic and the inelastic mean free paths can be realized / 63, 64 /. At low temperatures carrier transport through such short narrow constrictions can be entirely ballistic, i.e. without any scattering, and the conductance due to each 1D subband is quantized by 2e$^2$/h. The quantized energy levels are observed when the width of the constriction is changed by varying the gate voltage so that a sequence of subbands pass through the Fermi energy and the number of subbands contributing to conduction changes. As a result, the conductance decreases in steps of $2e^2/h$ as the channel width is narrowed. In addition, the resistance is independent of the length of the device provided the transport is ballistic.

If electron waveguides (i.e. semiconductor wires) are made longer than the electron mean free path, transport is no longer ballistic, but quantum effects are still observable. We have recently realized tunable 1D electron channels by directly written focussed ion beams (FIB) / 65 /. The striking new aspect of this "In-Plane-Gated (IPG)" transistor is that the confining electric field is <u>parallel</u> to the 2D electron gas at the heterointerface, and the distorted insulating region written by FIB acts as a dielectric. Both the conducting channel and the gates are thus formed by the 2D carrier gas in the same plane. At low temperature the IPG structure exhibits ballistic transport properties with a quantization of the conductance in multiples of $2e^2/h$. At room temperature the IPG structure operates as a field effect transistor (FET) with a very low capacitance due to its linear instead of planar gates. The FIB technique is particularly useful for the fabrication of low-dimensional semiconductors, because conducting and nonconducting regions in the material are defined directly and the resist exposure step is thus bypassed. Recently, Tarucha et al. / 66 / have confined the lateral dimensions of $GaAs/Al_xGa_{1-x}As$ double-barrier diodes by means of Ga FIB implantation. These diodes with a restricted lateral dimension exhibit a series of resonant-tunneling current peaks originating from the laterally confined 1D level, superimposed on the ground state confined by the heterojunction.

Using lithographically defined deep-mesa-etching of modulation doped $GaAs/Al_xGa_{1-x}As$ heterostructures and undoped quantum wells we have produced arrays of parallel quantum wires / 67 / and of quantum dots / 68 /. The periodic resist stripes or dots were prepared by holographic lithography. With reactive ion etching (RIE) in a $SiCl_4$ plasma rectangular groves of controlled depth were then etched through the multilayer structure. The crucial point of this process is that surface states on the etched sidewalls cause depletion of free carriers of different width, reducing the width of the wires or of the dots. In undoped $GaAs/Al_xGa_{1-x}As$ quantum wires we have recently observed 1D excitons / 69 /. The 1D character manifests itself in a strong polarization dependence and in an enhancement of the excitonic binding energy. In modulation-doped quantum wires, which represent extended 1D electronic systems, the potential, the energy levels, the number of electrons and the "active" wire width, were determined by transport measurements and far infrared (FIR) spectroscopy, both in magnetic fields / 67 /. Of importance is that collective phenomena strongly affect the electronic properties of this periodic array of quantum wires. Using the mesa-etching process discussed before we have also produced quantum dots in a periodic array / 68 /, which contain only 25 electrons per dot on discrete energy levels separated by 1 to 2 meV. For the first time we have thus realized artificial semiconductor "atoms", having an extremely large spatial extent. For reproducible fabrication of these low-dimensional semiconductor structures the pattern transfer techniques are often a more serious limitation than nanolithography. Therefore, we need to develop new methods of damage-free etching processes compatible with linewidths of less than 100 nm. Another approach to the lateral dimensional control is the use of molecular beam epitaxy of $GaAs/Al_xGa_{1-x}As$ heterostructures on slightly (2°) misoriented / 70 / or on carefully textured GaAs substrates / 71 /. In this case the quantum-wire active region is formed directly by the growth dynamics, eliminating the need for any intermediate patterning or processing that might introduce damage or contamination.

6. DEVICE APPLICATIONS OF CUSTOM-DESIGNED SEMICONDUCTOR MICROSTRUCTURES

The performance of conventional electronic and photonic devices can be improved considerably through the incorporation of custom-designed semiconductor microstructures, whose band gaps are engineered in an almost arbitrary and continuous way.One striking example for this application are optical detectors / 72 /. Using band-gap engineering, one can artificially tailor the ratio of the ionization coefficients and thus reduce excess noise. Another application of band-gap engineering to devices is the development of faster bipolar transistors, consisting basically of emitter, base, and collector. A device design that gives improved gain and response time makes use of a wide-gap emitter to form a heterostructure bipolar transistor (HBT). Ishibashi et al. / 73 / have developed a GaAs/Al$_x$Ga$_{1-x}$As HBT to produce the highest-speed small-scale functional digital integrated circuits. In these HBT structures a potential barrier in the valence band suppresses the back injection of holes in order to increase the current gain. In addition, the transport of electrons in the conduction band injected in the base is improved by launching carriers with some excess energies. This extreme nonequilibrium (i.e. ballistic) transport of electrons can be more fully exploited in Ga$_x$In$_{1-x}$As/InP HBTs, in which 165-GHz cutoff frequency has been measured /74/.

The two prototype devices, in which quantum phenomena arising from reduced dimensionality have been exploited widely for applications, are the quantum well laser / 75 / and the high electron mobility transistor / 76 /. The key aspects of quantum well (QW) lasers are (i) ultralow threshold (< 1 mA) currents, (ii) very narrow spectral linewidths, (iii) direct current modulation up to extremely high frequencies (> 1 Gbit/sec) (iv) low temperature sensitivity of the threshold current. These improvements rely in particular on the separate and precise control of electronic (gain) and optical confinement made possible by advanced epitaxial growth. Quantum well laser are especially well suited for monolithic integration with other active and passive components on a common substrate, because the absorption coefficient of the unexcited part of the QW structure is by a factor of five lower at the emitted wavelength than in conventional structures. Recently, surface-emitting lasers have gained renewed interest / 77 /. A 2D arrayed configuration of surface-emitting QW laser makes feasible high power capability, parallel optical processing, and vertical optical interconnection of circuit boards. Optimization of the preparation of optical microresonators and the application of strained Ga$_x$In$_{1-x}$As quantum wells enabled Scherer et al. / 78 / to fabricate surface-emitting laser-arrays with low threshold currents as well as all-optical microresonator switches with recovery times of 30 psec and controlling energies as low as 0.6 pJ. An ultimate improvement of laser characteristics with respect to threshold current, modulation bandwidth and frequency stability is expected from the incorporation of quantum wires or quantum dots in the active device region / 79 /.

The concept of modulation (or selective) doping of heterostructures and the formation of a 2D electron gas at the interface has been extensively exploited in the high electron mobility transistor (HEMT) / 76 /. Extremely low-noise GaAs/Al$_x$Ga$_{1-x}$As HEMT amplifiers for microwave applications have been commercially available by various manufacturers for several years. Multiwafer MBE systems are now used to produce modulation-doped heterostructures for the development of high-performance HEMT logic and memory LSI circuits / 80 /. A considerably higher 2D electron concentration and a high electron saturation velocity can be achieved when in conventional modulation-doped GaAs/Al$_x$Ga$_{1-x}$As heterostructures the GaAs is replaced by a strained Ga$_y$In$_{1-y}$As layer with y $\backsimeq$ 0.2. This modification leads to a strong increase of the conduction band discontinuity and hence allows a reduction of the Al mole fraction to eliminate the undesired persistent photoconductivity effect in Al$_x$Ga$_{1-x}$As with x > 0.2. As a result, greatly improved gain and noise characteristics have been observed in these pseudomorphic microwave HEMTs. The maximum available current-gain cutoff frequency todate is 270 GHz observed for a sub-0.1-μm gate-length device / 81 / . Another approach for extremely high-frequency and low-noise performance of HEMTs is the application of $Ga_x In_{1-x} As/Al_y In_{1-y} As$ heterostructures lattice matched to InP substrate. Chao et al. / 82 / reported cutoff-frequencies of more than 250 GHz for such transistors having gate lengths of 0.15 μm.

The combination of molecular beam epitaxy with nanoscale lateral patterning processes enables us to reduce the energy and length scales in custom-designed semiconductor structures such that macroscopic quantum phenomena are dominant. In these quantum structure devices (QSDs) the wave nature of the electron is fundamental to device performance. However, as yet the investigations of these quantum effects have mainly been done at very low temperature. For practical device applications it is indispensable to work at 77 K, i.e. semiconductors with lower effective carrier mass and minimum structural features of less than 50 nm are needed. In addition, the electron wave interference structures discussed in Sect. 5 operate under near-equilibrium conditions and do thus not offer any possibility for current gain. As a result, QSDs have to be developed which can be operated far away from equilibrium. Reed et al. / 83 / have laterally confined resonant tunneling heterostructures with a fabrication-imposed potential in order to produce quantum dots to be used for electronic transport studies. This QSD embeds a quasi-bound quantum dot (OD) between quantum wire (1D) contacts. In both the quantum dot and the quantum-wire contacts quantized electron wave functions are present. Arrays of such quantum dots have been fabricated which interact in a predictable way with their near neighbours. Moreover, these effects of interaction can be observed near room temperature and may thus form the basis to realize quantum coupled integrated circuits / 84 /. The IPG transistor described in Sect. 5    works at room temperature as a field-effect transistor with extremely low capacitance / 65 /. The transistor can operate in both 4-terminal and 3-terminal mode, allowing high integration without crossovers in one single fabrication step. For application in ultrafast logical and linear devices it is important to note that the new IPG structures are fabricated exclusively by writing FIB lines on as-grown heterostructures without any resist, mask, thermal, or chemical processing. With the present standard laboratory FIB systems it is possible to write as many as $10^6$ transistors in less than 10 seconds.

## 7. CONCLUDING REMARKS

The application of beam epitaxy to spatially resolved materials synthesis makes feasible the routine fabrication of a large variety of microscopically structured semiconductors exhibiting a (periodic) modulation in chemical composition perpendicular to the crystal surface down to atomic dimensions. In these artificially layered quasi-2D semiconductors the motion of free carriers is quantized within the layer plane. The reproducible fabrication of low-dimensional semiconductors having quasi-1D and quasi-OD electronic properties remains one of the major challenges in spatially resolved materials synthesis and, moreover, in the entire field of microstructure materials science. Methods must be developed which enable the removal of materials atom-by-atom in well-defined spatial and geometrical arrangements without causing damage to the crystal surface. The techniques described in Sect. 5 are ultimately not the best choice for the fabrication of quantum wires and quantum dots. The search for methods and techniques to manipulate the atoms in a crystal one-by-one (growth, removal, displacement) should be extended also to other solid materials. The most promising approaches for the direct synthesis of quantum wires and quantum dots include semiconductor microcrystals grown in dielectric media / 85 /, the selective coordinating epitaxy of mixed-valence metal compounds / 86 /, the matrix isolation of clusters / 87 /, the positioning of single atoms or molecules with a scanning tunneling microscope / 88 /, and the host/guest chemistry, as for example binary semiconductor clusters in cages of a zeolite / 89 /. In the latter method the host can even provide a three-dimensional periodicity to form a "supra-molecular" composition and hence the overall quantum lattice.

## ACKNOWLEDGEMENT

This work was sponsored by the Bundesministerium für Forschung und Technologie of the Federal Republic of Germany.

## REFERENCES

/ 1 / For a survey see: Proc.3rd Int. Conf. Modulated Semicond. Structures [MSS-3], Eds. A. Raymond and P. Voisin (Editions de Physique, Les Ulis, 1987) J. Physique 48, Colloque C5, C5/1 - C5/635 (1987); Proc. 4th Int. Conf. Modulated Semicond. Structures [MSS-4], Eds. L.L. Chang, R. Merlin, and D.C. Tsui (North-Holland, Amsterdam, 1990) Surf. Sci. 228, 1 - 578 (1990)

/ 2 / H. Sakaki, Proc. Int. Symp. Foundations Quantum Mechanics, Adv. Microfabr. Microstruct. Phys., Eds. S. Kamefuchi, H. Ezawa, Y. Murayama, M. Namiki, S. Nomura, Y. Ohnuki, and T. Yajima (Phys. Soc. Jpn., Tokyo, 1984) p. 94; F. Capasso, Science 235, 172 (1987)

/ 3 / K. Ploog, Angew. Chem. Int. Ed. Engl. 27, 593 (1988)

/ 4 / L.L. Chang and K. Ploog, Eds., Molecular Beam Epitaxy and Heterostructures (Martinus Nijhoff, Dordrecht, 1985) NATO Adv. Sci. Inst. Ser. E 87, 1 - 719 (1985)

/ 5 / G.B. Stringfellow, Organometallic Vapor Phase Epitaxy: Theory and Practice (Academic Press, Boston, 1989)

/ 6 / B.A. Joyce, Rep. Progr. Phys. 48, 1637 (1985)

/ 7 / M.B. Panish and H. Temkin, Annu. Rev. Mater. Sci. 19, 209 (1989)

/ 8 / W.T. Tsang, in VLSI Electronics : Microstructure Science, Ed. N.G. Einspruch (Academic Press, New York, 1989) Vol. 21, p. 255

/ 9 / S. Dushman, in Scientific Foundations of Vacuum Technique, Ed. J.M. Laffarty (John Wiley, New York, 1962) p. 80

/ 10 / For a recent survey see: Proc. 2nd Int. Conf. Chemical Beam Epitaxy 1989 [CBE-2], J. Cryst. Growth 105, 1 - 398 (1990)

/ 11 / A.Y. Cho and J.R. Arthur, Progr. Solid State Chem. 10, 157 (1975); A.C. Gossard, Treat. Mater. Sci. Technol. 24, 13 (1981)

/ 12 / For a recent survey see: Proc. 5th Int. Conf. Molecular Beam Epitaxy [MBE-V], Eds. Y. Shiraki and H. Sakaki (North-Holland, Amsterdam, 1989) J. Cryst. Growth 95, 1 - 637 (1989)

/ 13 / C.T. Foxon and B.A. Joyce, Current Topics Mater. Sci. 7, 1 (1981)

/ 14 / M. Ilegems, J. Appl. Phys. 48, 1278 (1977)

/ 15 / D.L. Miller and P.M. Asbeck, J. Appl. Phys. 57, 1816 (1985)

/ 16 / E. Nottenburg, H.J. Bühlmann, M. Frei, and M. Ilegems, Appl. Phys. Lett. 44, 71 (1984)

/ 17 / J. Maguire, R. Murray, R.C. Newman, R.B. Beal, and J.J. Harris, Appl. Phys. Lett. 50, 516 (1987)

/ 18 / L. Gonzales, J.B. Clegg, D. Hilton, J.P. Gowers, C.T. Foxon, and B.A. Joyce, Appl. Phys. A 41, 237 (1986)

/ 19 / W.I. Wang, Surf. Sci. 174, 31 (1986); H. Nobuhara, O. Wada, and T. Fujii, Electron. Lett. 23, 35 (1987)

/ 20 / E.H.C. Parker, Ed., The Technology and Physics of Molecular Beam

Epitaxy (Plenum Press, New York, 1985) 1 - 686; M.A. Herman and
H. Sitter, Molecular Beam Epitaxy, Fundamentals and Current Status
(Springer-Verlag, Berlin, 1989) Springer Ser. Mater. Sci. 7, 1 -
382 (1989)

/ 21 / A.Y. Cho, J. Appl. Phys. 42, 2074 (1971)

/ 22 / T. Sakamoto, H. Funabashi, K. Ohta, T. Nakagawa, N.J. Kawai, T.
Kojima, and K. Bando, Superlatt. Microstruct. 1, 347 (1985)

/ 23 / B.A. Joyce, P.J. Dobson, J.H. Neave, K. Woodbridge, J. Zhang,
P.K. Larsen, and B. Bölger, Surf. Sci. 168, 423 (1986)

/ 24 / B.A. Joyce, J. Zhang, J.H. Neave, and P.J. Dobson, Appl. Phys.
A 45, 255 (1988)

/ 25 / D.E. Aspnes, IEEE J. Quantum Electron. QE-25, 1056 (1989)

/ 26 / M.B. Panish, J. Electrochem. Soc. 127, 2729 (1980)

/ 27 / E. Veuhoff, W. Pletschen, P. Balk, and H. Lüth, J. Cryst. Growth
55, 30 (1981)

/ 28 / A. Robertson, T.H. Chiu, W.T. Tsang, and J.E. Cunningham, J. Appl.
Phys. 64, 877 (1988); D.A. Andrews and G.J. Davies, J. Appl. Phys.
67, 3187 (1990)

/ 29 / Y. Horikoshi and M. Kawashima, J. Cryst. Growth 95, 17 (1989)

/ 30 / F. Briones, L. Gonzales, and A. Ruiz, Appl. Phys. A 49, 729 (1989)

/ 31 / L. Esaki and R. Tsu, IBM J. Res. Develop. 14, 61 (1970)

/ 32 / B. Hamilton and A.R. Peaker, Prog. Cryst. Growth Character. 19,
51 (1989)

/ 33 / K. Ploog, Phys. Scr. T 19, 136 (1987); R. Cingolani, K. Ploog, L.
Baldassarre, M. Ferrara, M. Lugara, and C. Moro, Appl. Phys. A 50,
189 (1990)

/ 34 / K. Ploog, M. Hauser, and A. Fischer, Appl. Phys. A 45, 233 (1988)

/ 35 / E.O. Göbel, R. Fischer, G. Peter, W.W. Rühle, J. Nagle, and K.Ploog,
in Optical Switching in Low-dimensional Systems, Eds. H. Haug and
L. Banyai (Plenum Press, New York, 1989) NATO Adv. Sci. Inst. Ser.
B 194, 331 (1989)

/ 36 / G.H. Li, D.S. Jiang, H.X. Han, Z.P. Wang, and K. Ploog, Phys. Rev.
B 40, 10430 (1989)

/ 37 / J.B. Xia, Phys. Rev. B 38, 8358 (1988)

/ 38 / J. Feldmann, R. Sattmann, E.O. Göbel, J. Kuhl, J. Hebling, K.Ploog,
R. Muralidharan, P. Dawson, and C.T. Foxon, Phys. Rev. Lett. 61,
1892 (1989)

/ 39 / K. Ploog and R. Muralidharan, Def. Sci. J. 39, 367 (1989)

/ 40 / L.N. Pfeiffer, K.W. West, H.L. Störmer, and K.W. Baldwin, Mater.
Res. Soc. Symp. Proc. 145, 3 (1989)

/ 41 / E.F. Schubert, Y. Horikoshi, and K. Ploog, Phys. Rev. B 32, 1085
(1985); E.F. Schubert, J.E. Cunningham, and W.T. Tsang, Phys. Rev.
B 36, 1348 (1987); E.F. Schubert, J.E. Cunningham, and W.T. Tsang,
Appl. Phys. Lett. 51, 817 (1987)

/ 42 / L.L. Chang, L. Esaki, and R. Tsu, Appl. Phys. Lett. 24, 593 (1974)

/ 43 / H. Sakaki, T. Matsusue, and M. Tsuchiya, IEEE J. Quantum Electron.
QE-25, 2498 (1989)

/ 44 / T.P.E. Broekaert, W. Lee, and C.G. Fonstad, Appl. Phys. Lett. 53,
1545 (1988)

/ 45 / E.R. Brown, T.C.L.G. Sollner, C.D. Parker, W.D. Goodhue, and C.L.
Chen, Appl. Phys. Lett. 55, 1777 (1989)

/ 46 / J.F. Whittaker, G.A. Mouron, T.C.L.G. Sollner, and W.D. Goodhue,
Appl. Phys. Lett. 53, 385 (1988)

/ 47 / M.A. Reed, W.R. Frensley, R.J. Matyi, J.N. Randall, and A.C. Sea-
baugh, Appl. Phys. Lett. 54, 1034 (1989)

/ 48 / F. Capasso, S. Sen, F. Beltram, L. Lunardi, A. Vengurlekar, P.R.
Smith, N.J. Shah, R. Malik, and A.Y. Cho, IEEE Trans. Electron Devices ED-36, 2065 (1989)

/ 49 / R.F. Kazarinov and R.A. Suris, Fiz. Tekh. Poluprovodn. 5, 797 (1971)

/ 50 / S. Tarucha and K. Ploog, Phys. Rev. B 38, 4198 (1988)

/ 51 / S. Tarucha and K. Ploog, Phys. Rev. B 39, 5353 (1989)

/ 52 / J. Bleuse, G. Bastard, and P. Voisin, Phys. Rev. Lett. $\underline{60}$, 220 (1988)

/ 53 / E.E. Mendez, F. Agullo -Rueda, and J.M. Hong, Phys. Rev. Lett. $\underline{60}$, 2426 (1988)

/ 54 / K. Fujiwara, H. Schneider, R. Cingolani, and K. Ploog, Solid State Commun. $\underline{72}$, 935 (1989)

/ 55 / H. Schneider, K. Fujiwara, H.T. Grahn, K. von Klitzing, and K.Ploog, Appl. Phys. Lett. $\underline{56}$, 605 (1990)

/ 56 / J.W. Mathews and A.E. Blakeslee, J. Cryst. Growth $\underline{27}$, 118 (1974)

/ 57 / G.C. Osbourn, J. Vac. Sci. Technol. $\underline{A\ 3}$, 826 (1985)

/ 58 / I.J. Fritz and J.E. Schirber, Cryst. Prop. Prep. $\underline{21}$, 83 (1989) [Trans Tech Publ., Zürich, 1989]

/ 59 / E.P. O'Reilly, Semicond. Sci. Technol. $\underline{4}$, 121 (1989)

/ 60 / D. Gershoni and H. Temkin, J. Lumin. $\underline{44}$, 381 (1989)

/ 61 / O. Brandt, L. Tapfer, R. Cingolani, K. Ploog, M. Hohenstein, and F. Philipp, Phys. Rev. $\underline{B\ 41}$, 12599 (1990)

/ 62 / R. Cingolani, O. Brandt, L. Tapfer, G. Scarmacio, G.C. La Rocca, and K. Ploog, Phys. Rev. $\underline{B\ 42}$, 3209 (1990)

/ 63 / B.J. van Wees, H. van Houten, C.W.J. Beenakker, J.G. Williamson, L.P. Kouwenhoven, D. van der Marel, and C.T. Foxon, Phys. Rev. Lett. $\underline{60}$, 848 (1988)

/ 64 / D.A. Wharam, T.J. Thornton, R. Newbury, M. Pepper, H. Ahmed, J.E.F. Frost, D.G. Hasko, D.C. Peacock, D.A. Ritchie, and G.A.C. Jones, J. Phys. $\underline{C\ 21}$, L 209 (1988)

/ 65 / A.D. Wieck and K. Ploog, Appl. Phys. Lett. $\underline{56}$, 928 (1990)

/ 66 / S. Tarucha, Y. Hirayama, T. Saku, and T. Kimura, Phys. Rev. $\underline{B\ 41}$, 5492 (1990)

/ 67 / D. Heitmann, T. Demel, P. Grambow, and K. Ploog, in Advances in Solid State Physics, Ed. U. Rössler (Vieweg, Braunschweig, 1989) Vol. 29, p. 285

/ 68 / T. Demel, D. Heitmann, P. Grambow, and K. Ploog, Phys. Rev. Lett. $\underline{64}$, 788 (1990)

/ 69 / M. Kohl, D. Heitmann, P. Grambow, and K. Ploog, Phys. Rev. Lett. $\underline{63}$, 2124 (1989)

/ 70 / T. Fukui and H. Saito, J. Vac. Sci. Technol. $\underline{B\ 6}$, 1373 (1988); M. Tsuchiya, P.M. Petroff, and L.A. Coldren, Appl. Phys. Lett. $\underline{54}$, 1690 (1989)

/ 71 / E. Kapon, S. Simhony, R. Bhat, and D.M. Hwang, Appl. Phys. Lett. $\underline{55}$, 2715 (1989); K. Kojima, K. Mitsunaga, and K. Kyuma, Appl. Phys. Lett. $\underline{56}$, 154 (1990)

/ 72 / F. Capasso, in Physics and Applications of Quantum Wells and Super-lattices, Eds. E.E. Mendez and K. von Klitzing (Plenum Press, New York, 1987) NATO Adv. Sci. Inst. Ser. $\underline{B\ 170}$, 377 (1987)

/ 73 / T. Ishibashi and Y. Yamauchi, IEEE Trans. Electron Devices $\underline{ED-35}$, 401 (1988); Y. Yamauchi, K. Nagata, and T. Ishibashi, Inst. Phys. Conf. Ser. $\underline{91}$, 693 (1988)

/ 74 / Y.K. Chen, R.N. Nottenburg, M.B. Panish, R.A. Hamm, and D.A.Humphrey, IEEE Electron Device Lett. $\underline{EDL-10}$, 470 (1989)

/ 75 / L.C. Chin and A. Yariv, J. Lumin. $\underline{30}$, 551 (1985)

/ 76 / M. Abe, T. Mimura, K. Nishiuchi, N. Yokoyama, in VLSI Electronics: Microstructure Science, Ed. N.G. Einspruch (Academic Press, New York, 1985) Vol. 11, p. 333; T.J. Drummond, W.T. Masselink, H.Morkoc, Proc. IEEE $\underline{74}$, 773 (1986)

/ 77 / K. Iga, F. Koyama, and S. Kinoshita, J. Vac. Sci. Technol. $\underline{A\ 7}$, 842 (1989)

/ 78 / A. Scherer, J.L. Lewell, Y.H. Lee, J.P. Harbison, and L.T. Florez, Appl. Phys. Lett. $\underline{55}$, 2724 (1989)

/ 79 / A. Yariv. Appl. Phys. Lett. $\underline{53}$, 1033 (1988)

/ 80 / K. Kondo, J. Saito, T. Igarashi, K. Nanbu, and T. Ishikawa, J. Cryst. Growth $\underline{95}$, 309 (1989); T. Sonoda, M. Ito, M. Kobiki, K. Hayashi,

S. Takamiya, and S. Mitsui, J. Cryst. Growth <u>95</u>, 317 (1989)

/ 81 /  P.C. Chao, M.S. Shur, R.C. Tiberio, K.H.G. Duh, P.M. Smith,
J.M. Ballingall, P. Ho, and A.A. Jabra, IEEE Trans. Electron
Devices <u>ED-36</u>, 461 (1989)

/ 82 /  P.C. Chao, A.J. Tessmer, K.H.G. Duh, P. Ho, M.Y. Kao, P.M. Smith,
J.M. Ballingall, S.M.J. Liu, and A.A. Jabra, IEEE Electron Device
Lett. <u>EDL-11</u>, 59 (1990)

/ 83 /  M.A. Reed, J.M. Randall, J.H. Luscombe, W.R. Frensley, R.J.Aggarwal,
R.J. Matyi, T.M. Moore, and A.E. Wetzel, in <u>Advances in Solid State
Physics</u>. Ed. U. Rössler (Vieweg, Braunschweig, 1989) Vol. 29,
p. 267

/ 84 /  J.M. Randall, M.A. Reed, and G.A. Frazier, J. Vac. Sci. Technol.
<u>B 7</u>, 1398 (1989).

/ 85 /  N. Chestnoy, R. Hull, and L.E. Brus, J. Chem. Phys. <u>85</u>, 2237 (1986);
A.I. Ekimov, I.A. Kudryavtsev, M.G. Ivanov, and A.L. Efros,
J. Lumin. <u>46</u>, 83 (1990)

/ 86 /  K. Takahashi, H. Tanino, and T. Yao, Jpn. J. Appl. Phys. 26, L97
(1987)

/ 87 /  T.P. Martin, Angew. Chem. Int. Ed. Engl. <u>25</u>, 197 (1986)

/ 88 /  D.M. Eigler and E.K. Schweitzer, Nature <u>344</u>, 524 (1990)

/ 89 /  G.D. Stucky and J.E. MacDougall, Science <u>247</u>, 669 (1990)

# METAL ORGANIC VAPOUR PHASE EPITAXY FOR THE GROWTH OF SEMICONDUCTOR STRUCTURES AND STRAINED LAYERS

M.R. Leys

Semiconductor Physics
Technical University of Eindhoven
Box 513
5600 MB Eindhoven, The Netherlands

## INTRODUCTION

The technological development of semiconductor materials started in the period following the second world war. In the electronics industry, the first transistors were fabricated from germanium, later from silicon. It was soon realized that also the $A^{III}\text{-}B^{V}$ or $A^{II}\text{-}B^{VI}$ materials (most often simply termed III-V or II-VI materials) exhibited semiconductive behaviour. The energy difference between the valence band and the conduction band made them candidates for electronic devices which can absorb or emit phonons over a range of frequencies (wavelengths). Direct bandgap materials such as gallium arsenide (GaAs) were suitable for devices in which efficient electron–hole recombinations could take place and high efficiency light emitting devices were a possibility. Stimulated emission was first demonstrated in 1970 with the preparation of the single heterojunction and the double heterojunction laser diodes. These devices are multiple layer structures with a thin waveguide region contained between layers of larger bandgap and different refractive index (for confinement of carriers and radiation, respectively, in the active region). A basic laser diode chip consists of two parallel facets, (110) planes, which are prepared by cleavage and act as mirrors. The Fabry-Perot cavity is defined by these two parallel facets and the passive (cladding) layers. In the longitudinal direction current definition is by mesa etching and/or stripe-contact metallization.

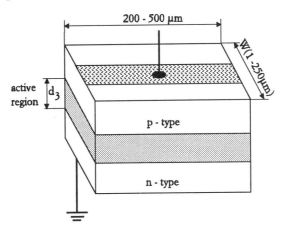

Fig. 1.    Schematic illustration of the double heterojunction laser.

The first succesfull heterojunction laser consisted of a multiple layer structure and was based on the (Al,Ga)As alloy. GaAs and AlAs have approximately equal lattice constants (5.635 Å and 5.661 Å, respectively), but a rather large difference in bandgap (1.42 eV and 2.16 eV) and refractive indices (3.55 and 2.97, respectively, with $\lambda = 0.9$ $\mu$m). This implies that a laser structure such as shown in fig. 1 can be fabricated from (Al,Ga)As as a single crystalline structure, with perfect continuity of the crystal through all layers, and also perfect lattice matching to a GaAs substrate.

In fig. 2, a diagram is given of the bandgap of III-V compounds as function of lattice parameter.

The indirect bandgap materials are indicated in the upper left. As substrate materials, GaP, GaAs, InP, InAs, and InSb are obtainable. In order to obtain crystalline

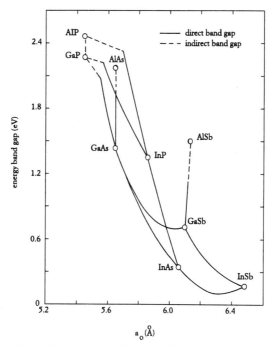

Fig. 2.  Relation between bandgap and lattice constant for various III-V compounds.

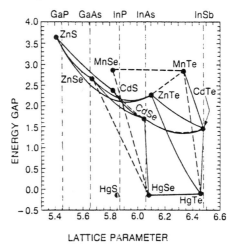

Fig. 3.  Relation between bandgap and lattice constant for various II-VI compounds.

continuity through all interfaces of a multiple layer structure grown on one of the previously mentioned substrates, it is required to perfectly match the lattice constants of each single layer to the lattice constant of the substrate. With exception of the (Al,Ga)As and (Al,Ga)P systems, this lattice-match requirement neccessitates a high degree of control on the composition of the layers to be grown. In figure 3, the bandgap versus lattice constant of the II–VI materials is shown. Also indicated are the lattice constants of the commercially available substrates.

Clearly, the II–VI materials cover a much broader bandgap range than the III–V compounds. However, the growth of II–VI materials has not (yet) reached the level of control that has been achieved in the III–V materials. Notably, problems are encountered with an appropriate choice of substrates and in obtaining a sufficiently high purity level of the layers.

In 1969 a novel growth technique for preparing III-V and II-VI epitaxial layers was demonstrated [1]. The technique made use of hydrides for the group V component and metal organic compounds as group III sources. It was named Metal Organic Vapour Phase Epitaxy (MOVPE). In this first paper the growth of galliumarsenide layers was described using the metal organic compounds trimethylgallium ($Ga(CH_3)_3$, TMGa) and the hydride, arsine ($AsH_3$) transported by a carrier gas to a heated surface. The overall chemical reaction was formulated as follows:

$$Ga(CH_3)_3 + AsH_3 \xrightarrow{\text{heat}} GaAs + 3CH_4$$

In a subsequent series of articles Manasevit proved the possibilities of this growth technique by growing a very complete series of the III–V compounds: GaP was grown by using TMGa and phosphine ($PH_3$), the ternary material $Ga(As,P)$ by introducing both phosphine and arsine and growth of $Ga(As,Sb)$ was achieved by introducing stibine ($SbH_3$) together with arsine. Also, AlAs was grown, by making use of trimethylaluminium ($Al_2(CH_3)_6$), as well as the compounds GaN and AlN, by introducing ammonia ($NH_3$) into the reactor. Finally, the indium containing compounds InAs, $(Ga,In)As$, $In(As,P)$ and InP were also grown. It appeared, that also II–VI materials, such as $Zn(S,Se)$ and $(Cd,Hg)Te$ could be grown when using appropriate metal organic compounds and hydrides.

A schematic drawing of the equipment required for the growth of, for example, a GaAs/(Al,Ga)As laser diode structure is given in figure 4. As carrier gas for the reagents most often hydrogen is used, purified by a palladium diffuser. The amount of hydrogen passed through each gas line is regulated by electronic mass flow controllers. The metal–organic compounds (here, trimethylgallium (TMGa), trimethylaluminium (TMAl), and diethylzinc (DEZn) for p–type doping) are contained in stainless steel cylinders which are kept in a thermostated bath to give the desired vapour pressure. Their vapours are then transported to a gas switching manifold where the component is made to flow either into the reactor or, when (momentarily) not required for growth, is vented to the exhaust line. A similar type of switching is carried out with the lines transporting the gases arsine ($AsH_3$) and silane ($SiH_4$, for n–type doping). The susceptor with the substrate is held within a quartz glass reactor and growth takes place at a temperature of $\sim 700^0C$. The partial pressure (concentration) of the group III component is mostly of the order of $10^{-4}$ atm.; the group V component is always present in excess.

The technique of MOVPE has, because of its similarity with the extremely important silicon deposition process, a potential towards mass production of optoelectronic devices. In fact, the possibility of MOVPE to produce low threshold double heterojunction lasers as early as in 1977 [2] has meant an important stimulus towards application of the technique. By making use of low pressure MOVPE, impressive results were obtained in the $InP-(In,Ga)(As,P)-InP$ materials system, with lasers for optical communication emitting at 1.26 and 1.55 $\mu m$ with a low threshold current [3].

Active research on the possibilitities of MOVPE is taking place in a number of fields. For large scale production, emphasis is on the design of the reactors and on the understanding of the behaviour of the carrier gas flow through the reactor tube. Furthermore, research is being carried out on understanding the chemistry of the process. New metal organic sources and alternatives to notably the hydrides (arsine, phosphine) are being developed, with emphasis on high (electronic grade) purity. Also, investigations are carried out to understand the fundamental mechanisms of the process: the way in

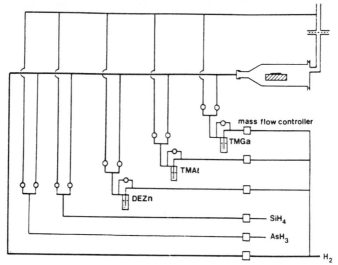

Fig. 4. Schematic diagram of a MOVPE growth system. The metal organic compounds trimethylgallium (TMGa), trimethylaluminium (TMAl) and diethylzinc (DEZn) are liquids and are stored in stainless steel bubblers. The gases arsine ($AsH_3$) and silane ($SiH_4$) are stored in cylinders. $H_2$ is used as carrier gas.

which the precursor molecules (metal organics and hydrides) decompose and arrange on the growing surface so that smooth, epitaxial layers result.

In this paper, we first review the aspects of hydrodynamics and reactor design. Then we will discuss the starting compounds used in MOVPE and the chemical reactions of trimethylgallium, arsine and aluminium precursors. After that, the growth mechanism of GaAs epitaxial layers is reviewed. The final chapter of this paper is devoted to layers grown to a lattice constant which differs from the lattice constant of the substrate, the so–called strained layers.

## REACTOR DESIGN/FLOW DYNAMICS

In principle, the design for a MOVPE reactor is extremely simple: a cold wall tube (either air or water cooled) made of quartz glass containing within it a susceptor made from graphite. The susceptor is heated to the desired growth temperature by r.f. induction. Alternatively, one may use infrared heating from quartz–halogen lamps. A great variety of reactor designs have been proposed by the various workers in the field of MOVPE. A discussion on the many different reactor configurations has been given in a previous review paper on MOVPE [4]; here, we update this compendium by briefly discussing the two novel types which have appeared since publication of reference 4. These two new reactor types are schematically shown in figures 5c and d.

The initial work of Manasavit was carried out in a vertical reactor, as drawn in figure 5a. To obtain good homogeneity in layer thickness, rotation of the susceptor is required. The horizontal design (figure 5b) was introduced for MOVPE growth by Bass [5] and such a reactor operating at reduced pressure ( 100 Torr) was promoted by Duchemin [6]. To obtain a good homogeneity in layer thickness in a horizontal reactor the reactor geometry must be designed carefully. The planetary motion–reactor shown in figure 5c is a large–scale reactor specifically designed for the growth of III–V materials with emphasis on layer thickness homogeneity and the possibility to abruptly change the gas composition. The details of this reactor are described in [7]. It consists of a rotating main platform with, for each substrate, a satellite plateau which is suspended and rotates on a gas foil. The upper– and lower part of the reactor have independent temperature control and the metal organics and hydrides are introduced separately. The pulse reactor (figure 5d, reference [8]) has a completely different design philosophy than the above mentioned continuos–flow reactors. This reactor consists of a set of inlet pipettes of adjustable volume, valves, a substrate container and a valved pumping system. Operation is by inlet– followed by pump–pulses. Each inlet pulse is made to react to exhaustion, after which the reaction products are pumped out of the chamber before the next pulse of component(s) is let in. Thus, a high degree of efficiency can be expected as well as an accurate control over the layer thicknesses.

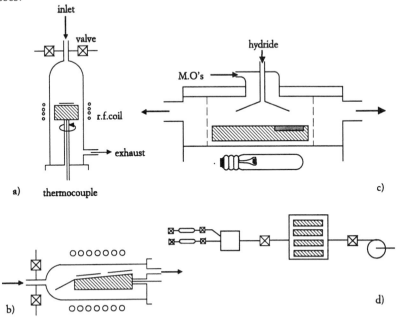

Fig. 5   Schematic diagrams of (a) Vertical reactor; (b) Horizontal reactor; (c) Planetary motionreactor; (d) Pulse reactor.

To evaluate the merits of the various reactor designs a knowledge of the behaviour of the carrier gas is required. The continous–flow MOVPE reactors are characterized by the fact that a gas flow with a velocity of ~ 30cm/sec is passed through the tube and that there is only one hot area in the system, this being the susceptor. These two facts mean that one has to deal with two types of fluid transport. In the first place, one has forced convection: the flow of carrier gas through the reactor on account of the pressure difference between inlet and exhaust. In most VPE reactors, this flow is expected to be of the laminary type, with the Reynolds number being below a value of 100. As the second type of fluid movement, one has to deal with free convection: flow which may occur due to differences in temperature (and thus in fluid density) in different regions in the system. Free convention is characterized by the Rayleigh number; above a certain critical value of this number the hot (less dense) fluid will start to rise and then overturn the upper, colder layer of the fluid. The tendency of a hot fluid to rise is termed "buoyancy".

In studies to describe and optimize the epitaxial growth of silicon in a horizontal reactor (as shown in figure 5b) the flow of the carrier gas was vizualized by making use of small particles of $TiO_2$ as markers of the streamline patterns [9]. From these studies, a model was developed in which two regions were distinguished in the reactor: adjacent to the susceptor one envisaged a slow moving (even stagnant) layer of gas over which a steep thermal gradient exists. In the region above this boundary layer, one has the free flowing layer, where the gas velocity is high and where temperature differences are less pronounced. In this simple model, the mass transfer controlled process could be described as a process where diffusion of components through a boundary layer was the rate limiting step for epitaxial layer growth. Variations of the growth rate over the susceptor could then be attributed to variations in the thickness of the boundary layer or to changes in the concentration gradient over this same layer.

Such variations are bound to occur close to the leading edge of the susceptor (where the boundary layer is still developing) and further along the length of the reactor, due to depletion of the gas phase.
However, complications in this simple picture arise from the possibility that fluid form the hot region close to the susceptor can rise up and disturb the uniformity of the flow. Transverse– and/or longitudinal rolls may occur in the horizontal reactor geometry by the influence of the free convection on the (in principle) laminary forced convection flow.The interactive behaviour of forced– and free convection can give rise to non–uniformities in layer thickness and can also give rise to delay times in the introduction or removal of components. Such transients in gas–switching can lead to non–abrupt compositional changes in the epitaxial structure. Computer calculations are required to optimize reactor design. These calculations involve solution of the Navier–Stokes equations in two– or three dimensions via the finite–element method to give trajectories of a fluid particle through a reactor tube. An example of such a fluid particle trajectory through a horizontal reactor is shown in figure 6, from reference [10].

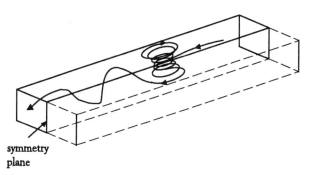

symmetry
plane

Fig. 6.    Fluid particle trace in three dimensional flow involving both a transverse roll (return flow) and longitudinal rolls. The reactor is symmetric with respect to the midplane. From reference 10.

In reference 11, by time dependent computer simulations, recirculation effects are calculated in relation to reactor geometry and operating presuure. It is shown that reciculation effects gives rise to trapping of components inside the reactor due to which the abruptness of interfaces between layers of different composition is negatively effected. Analytical models to describe the growth rate uniformity in a horizontal reactor and analytical models to assess the limiting factors which determine the interface abruptness in VPE reactors are given in [12] and [13], respectively.
The behaviour of the carrier gas flow through the VPE reactor depends on a number of factors. First, it will depend on the absolute flow rate as this determines the scale of the entrance effects. These entrance effects are more severe when sudden expansions or restrictions are present in the reactor tube. Next, the reactor pressure is of importance: reducing the pressure tends to smooth the streamline pattern to the ideal, laminar type. Finally, the choice of carrier gas (in combination with geometry) determines to what extent the free convection movement influences the forced flow. An ambient of a gas with a high density and a low heat conductivity (such as nitrogen or argon) gives rise to strong free convection motion.
From figure 5 it can be seen that both the reactor types a and b are geometries in which the free convection and forced convection movement are opposing, respectively perpendicular to each other. To be free of convective instability, these reactors are best operated with a high carrier gas flow rate or at reduced pressure. In fact, experiments on the flow visualization in the vertical reactor have shown that a pressure decrease down to ~ 200 Torr eliminates the free convection effects completely [14]. Therefore, there is a trend in reactor technology to operate at reduced pressure. Decreasing the pressure of operation of the reactor decreases the importance of the gas flow behaviour. In the proposed vacuum chemical epitaxy (VCE) reactor [15], the

carrier gas effects can be left out completely. Also in the case of chemical beam epitaxy (CBE) [16] there is no carrier gas to transport species towards the substrate surface. Alternatively, in the atmospheric MOVPE process, one can make maximal use of cooperative convective flows for transport of components in well designed reactors.

## CHEMISTRY

Most often in MOVPE, the methyl compounds of the group III metals are used: trimethylgallium, trimethylaluminium and trimethylindium. The compounds with the ethyl group as a ligand are also applied: triethylgallium, triethylaluminium and triethylindium. As source for the group V element, the use of hydrides is common practice: arsine for the growth of arsenides and phosphine for the growth of phosphides. For the growth of the nitrides, ammonia is used although also hydrazine ($N_2H_4$) has been proposed [17]. Antimodes are grown with stibine ($SbH_3$) but recently most often with trimethylantimonide [18]. At present, research is carried out to substitute the highly toxic hydrides with organo–V compounds. Thus, trimethylarsenic, triethylarsine and phenylarsine [19] have been investigated as alternatives to arsine. As alternatives to phospine, the compounds isobutylphosphine and tertiarybutylphosphine have been studied [20].

As dopants, a wide range of chemicals has been used. For p–type doping, or for the growth of II–VI compounds, the methyl compounds dimethylzinc [21] and dimethylcadmium [22] and the ethyl compound diethylzinc [21–23] can be used. Furthermore, diethylberylium [24] and bis–cyclopentadienyl magnesium exist as p–type dopant sources. For n–type doping in III–V materials, most often hydrides are used: hydrogensulfide and hydrogenselenide [25], silane [6] and disilane ($Si_2H_6$) [26]. Also germane ($GeH_4$) acts as an n–type dopant [6], [24]. Metal organic sources for n–type doping are tetramethyl– or tetraethyltin [27], [28]. Dimethyl–and diethyltellurium may be used as a tellurium source, both for n–type doping in III–V materials as well as in the growth of Tellurium–containing II–VI materials (see Ref. 29). For the growth of semi–insulating layers, iron can be incorporated via iron–olefin molecules [30], vanadium has been incoporated via triethoxyvanadyl ($VO(OC_2H_5)_3$ [31] and chromium doping has been achieved with hexacarbonylchromium [32]. Finally, we mention that doping of e.g. InP with rare–earth elements is receiving some attention. For example, Ytterbium doping can be achieved by making use of a tri–isopropylcyclopentadienyl compound [33].

It is beyond the scope of this paper to treat the decomposition kinetics and the incorporation mechanisms of all these compounds. Instead, the discussion here will be limited to just a few compounds with relevance to the growth mechanism of epitaxial gallium arsenide layers.

## GALLIUM ARSENIDE GROWTH

Investigations on the surface reaction between TMGa and $AsH_3$ in the temperature range of 200 and 420$^0$C were carried out in 1977 [34]. It was then proposed that GaAs layer growth takes place via Langmuir–type adsorption of the complete metalorganic– and complete hydride–molecule on neighbouring surface sites. After adsorption of the two molecules on the surface, chemical reactions take place and methane is desorbed from the surface. Later investigations, made use of infrared spectroscopy to analyse gas samples taken from the MOVPE reactor operating under growth conditions [35]. These investigations revealed that trimethylgallium is a relatively unstable molecule and easily undergoes a decomposition reaction in the gas phase. The arsine molecule is relatively stable, but the decomposition of this molecule is enhanced by the presence of GaAs surfaces and by the products formed upon of TMGa pyrolysis. We make note here that unstable (short–lived) intermediates, or compounds with low vapour pressure, are not easily observed with ex–situ sampling techniques. In an extensive study of the epitaxial growth of GaAs between temperatures to 450 to 1050$^0$C it was shown that three distinct temperature regimes can be distinguished [36]. At temperatures below 650$^0$C the growth rate is strongly dependent on substrate temperature. This implies that chemical reactions, taking place on the surface or that

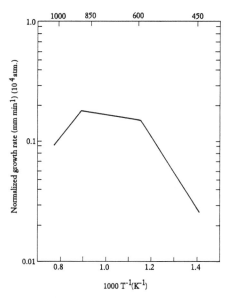

Fig. 7. Plot of the normalized growth rate of GaAs epitaxial layers as function of temperature. From reference 4.

desorption of products from the surface is the rate limiting step in the growth process. At temperatures above 750°C, the growth rate decreases with increasing temperature. This decrease in growth rate is caused by an increased desorption rate of gallium from the growing surface (in MBE terminology: due to an decrease in the sticking coefficient of gallium; in the low– and mid– temperature regimes the sticking coefficient of gallium is equal to 1, at higher temperatures the gallium sticking coefficient decreases). In the mid–temperature regime, from 650 to 750°C, the growth rate is independent of temperature and independent of the amount of arsine introduced into the reactor. Under these conditions, the growth rate is limited by mass–transfer (i.e. diffusion through the boundary layer adjacent to the hot susceptor surface) and dependent only on the arrival rate of gallium species. This implies that chemical reactions at the surface and incorporation of atoms in the crystal lattice are relatively fast steps in the overall process.

Recently, the results of calculations which include heat and mass transport, as well as reaction kinetics in the gas phase and on the (111) GaAs surface have become available [37]. Also, real in situ diagnostics have been carried out making use of infrared laser diode spectroscopy [38] and Raman spectroscopy [39]. Both the calculations as well as the measurements make it possible to quantitatively determine concentration profiles of all possible intermediate species which are present in the reactor during growth. Thus, the main gas phase reactions have been established as

$$Ga(CH_3)_3 \longrightarrow \ ^\cdot Ga(CH_3)_2 + \ ^\cdot CH_3 \qquad (1)$$

$$Ga(CH_3)_2 \longrightarrow \ :Ga–CH_3 + \ ^\cdot CH_3 \qquad (2)$$

$$^\cdot CH_3 + AsH_3 \longrightarrow CH_4 + \ ^\cdot AsH_2 \qquad (3)$$

The trimethylgallium molecule dissociates into a dimethylgallium radical and a methyl radical; dimethylgallium rapidly loses a second methyl group to form mono-methylgallium. The dots placed at the chemical species indicate the presence of a free, unpaired electron. In principle, a reaction:

$$^\cdot H + Ga(CH_3)_3 \longrightarrow \ ^\cdot Ga(CH_3)_2 + CH_4 \qquad (4)$$

can also be envisaged for the decomposition of trimethylgallium in hydrogen. This reaction (4) would take place in parallel to reaction (1), leading to an enhancement of the decomposition rate due to the presence of hydrogen.

76

The methyl radicals liberated by reactions (1) and (2) are able to extract hydrogen from arsine. The thus formed $\cdot AsH_2$ can undergo subsequent loss of hydrogen to give $:AsH$ and $As$ and can, eventually give the thermodynamically favoured species $As_4$. Under "normal" operating conditions of the MOVPE process, the equilibrium species are not formed due to the fact that the kinetic steps are relatively slow compared to the dwell times of the species in the hot zone of the reactor.

We consider the following reactions to be the most likely surface processes

$$GaCH_{3ads} + AsH_{ads} \longrightarrow GaAs + \underset{\underset{CH_{4g}}{\downarrow}}{\cdot CH_{3ads}} + \cdot H_{ads} \qquad (5)$$

and/or

$$GaCH_{3ads} + As_{ads} \longrightarrow GaAs + \underset{\underset{CH_{3g}}{\uparrow\downarrow}}{\cdot CH_{3ads}} \qquad (6)$$

Without $AsH_x$ species present, and with hydrogen

$$GaCH_{3ads} + H_2 \text{ (or } \cdot H) \longrightarrow Ga_{ads} + \cdot CH_{3ads} + (2) \underset{\underset{CH_{4g}}{\downarrow}}{\cdot H_{ads}} \qquad (7)$$

Note that monomethylgallium arrives as the group III species on the growing surface. When excess arsine (or arsenic) is present crystalline gallium arsenide is formed, according to, respectively, reactions (5) and (6).

However, when GaAs is grown with $Ga(CH_3)_3$ and $As_4$ from an effusion cell an UHV (MBE) system, $p^{++}$ type layers are obtained, with carbon levels up to $\sim 10^{20}$ cm$^{-3}$ [40]. Growth of GaAs layers using trimethylgallium and arsenic evaporated from a solid source in a hydrogen ambient, also gives rise to heavily p–type layers [41]. These results imply that monomethylgallium, adsorbed on a single crystal GaAs surface, can be made to lose methyl to a concentration of less than 0.1% under influence of arsenic species. However, the carbon levels obtained when using trimethylgallium in combination with $AsH_3$ are below the level of $10^{15}$ cm$^{-3}$. The decrease of carbon contamination by five orders of magnitude when $AsH_3$ is used instead of $As_4$ suggests that atomic hydrogen, present on a reactive surface, strongly aids the complete removal of the final methyl group from $GaCH_3$. In summary, the rate limiting step of $Ga(CH_3)_3$ decomposition is dissociation of the first methyl group. In MOVPE, this takes place as a homogeneous, gas phase reaction. The liberated methyl radicals remove a hydrogen atom from $AsH_3$ (reaction (3)) or, when no arsine is present, react with hydrogen carrier gas. In both cases, methane is formed. Most likely, the final methyl group from trimethylgallium is removed on the surface. It can be said that the growth of epitaxial GaAs layers by MOVPE in the mid–temperature range takes place uniquely via Equations (1)–(6) given above. We make note here that this mechanism of epitaxial GaAs growth is not significantly dependent on reactor design. Changes in the type of ambient (e.g. from $H_2$ to He) or changes in the total pressure (from 760 to approximately 20 torr) and changes in the dwell time of species in the reactor affect mainly the relative importance of the reactions (5) with respect to reaction (6).

## GROWTH MECHANISM OF $Al_xGa_{1-x}As$

For the growth of $Al_xGa_{1-x}As$ layers with MOVPE most often the combination trimethylgallium together with trimethylaluminium (TMAl) is used, with arsine as the group V source. With this combination, the concentration x in the epitaxial layer is linearly proportional to the concentration of TMAl in the reactor ambient. This implies, that the growth mechanism of AlAs, in the mid–temperature range, is very similar to the growth mechanism of GaAs: a growth rate determined only by the arrival rate of group III species on the substrate surface. A striking difference between AlAs and GaAs grown by MOVPE is the fact that, when pure AlAs is grown using TMAl and $AsH_3$, the level of carbon contamination can be as high as $5 \times 10^{17}$ cm$^{-3}$. This is three orders of magnitude higher than in GaAs layers, grown under identical growth conditions, using TMGa and $AsH_3$. This carbon incorporation in AlAs must be related to the use of the methyl–containing aluminium metal organic precursor as AlAs layers grown from triethylaluminium have a non–detectable carbon level [42], [43].

The actual decomposition mechanism of trimethylaluminium has up to now not yet been resolved and reliable values for the dissociation energies are not known. A review of properties of trimethyl– and triethylaluminium is given in references 44 and 45; the properties are summarized below.

Trimethylaluminium is a dimeric molecule in the solid, liquid and vapour phase up to about $260^0$C. Thus, in the boundary layer region of the MOVPE reactor, TMAl is present as a monomer. As such, it could undergo a pyrolysis reaction similar to TMGa, i.e.

$$Al(CH_3)_3 \longrightarrow {}^{\cdot}Al(CH_3)_2 + {}^{\cdot}CH_3 \tag{8}$$

$$^{\cdot}Al(CH_3)_2 \longrightarrow {:}Al{-}CH_3 + {}^{\cdot}CH_3 \tag{9}$$

Formation of epitaxial AlAs layers could then take place via reactions analogous to (5) and (6). Due to the uncertainty in the dissociation energies for methyl removal, it is unclear whether reactions (8) and (9) take place in the gas phase or on the surface.

A second problem is that one cannot, from the previously published work on TMAl pyrolysis, give the chemical composition of the final product of TMAl pyrolysis. Also, the effect of hydrogen on both the reaction rate and on the composition of the final product of TMAl pyrolysis is unknown. The high carbon content in AlAs layers and also the fact that pyrolysis of TMAl in hydrogen gives rise to a final product containing up to 50 atomic percent carbon [46], [47] indicates that severe kinetic limitations exist for removal of the final methyl group from TMAl.

In [47], the pyrolysis of TMAl was proposed to proceed via the following mechanism:

$$Al_2(CH_3)_6 \longrightarrow 2Al(CH_3)_3 \tag{10}$$

$$Al(CH_3)_3 \longrightarrow {}^{\cdot}Al(CH_3)_2 + {}^{\cdot}CH_3 \longrightarrow CH_3Al{=}CH_2 + CH_4 \tag{11}$$

$$CH_3Al{=}CH_2 \longrightarrow Al{\equiv}CH + CH_4 \tag{12}$$

The attractiveness of these equations lies in the fact that formation of a final product with a strong aluminium to carbon bond can be explained. This then illustrates the difference between the final product of TMAl pyrolysis with respect to the final product of TMGa pyrolysis (Ga–CH$_3$). This, in turn, illustrates the cause of the high background carbon contamination in AlAs layers as compared to GaAs layers grown with the methyl containing metal organic molecules.


## STRAINED LAYERS

In general, multiple layer semiconductor devices consist of layers with equal lattice constant. This is readily achieved in the GaAs–AlAs system, as the lattice constants of these two materials do not differ much. For GaAs the lattice constant at room temperature, $a_o = 5.653$ Å; for AlAs $a_o = 5.661$ Å, so that the relative mismatch $f = \dfrac{\Delta a}{a} = 1.4 \cdot 10^{-3}$. In the $In_{1-x}Ga_x As_yP_{1-y}$–InP system, lattice matching of the ternary and/or quaternary layer to the InP substrate is achieved by carefull control of the values of x and y. In that way, values of $\dfrac{\Delta a}{a} = 10^{-3}$ can also be achieved in these structures. Lattice matching is considered necessary in order to avoid the occurrence of misfit dislocations in the epitaxial structures. Dislocations negatively affect the operating lifetime of the semiconductor devices.

78

With the advent of thin-layer growth technology, obtained with the VPE and MBE processes, it became possible to fabricate epitaxial structures in which the total amount of mismatch could be taken up by elastic deformation of the epitaxial layer, thus preventing the formation of misfit dislocations. The growth of such strained--layers has since then received considerable attention. First of all, lattice strain alters the band-gap structure of the semiconductor material, thereby creating a larger degree of flexibility for the materials scientist. Next, the mechanism of formation of misfit dislocations has become an interesting field of research. This paragraph deals mainly with the latter topic. For the possibilities of strain as a additional tool in materials science, see e.g. [48], for a detailed discussion of the effects of strain on the bandgap structure, we refer to [49].

## NUCLEATION OF MISFIT DISLOCATIONS

As has been mentioned in the introduction, the advent of thin-layer growth technology has provided the breakthrough for thorough investigation of the possibilities and problems of strained layer structures. The most well-known discussion on this topic is given in [50]. Here, it is proposed that the amount of mismatch $f\left(=\frac{\Delta a}{a}\right)$ is taken up by elastic strain ($\epsilon$) or by misfit dislocations ($\delta$), so that $f = \epsilon + \delta$. Here $\delta = b_\parallel/d$ while $b_\parallel$ stands for the projection of the Burgers vector of the dislocation, parallel to the interface and d is the average separation between dislocations.

Matthews and Blakeslee [50] proposed a thermodynamic model, the so-called mechanical equilibrium theory, in which the total amount of elastic strain energy incorporated in the lattice is balanced with the energy stored in a square grid of misfit dislocations. When the force on a dislocation line (e.g. a threading dislocation originating from the substrate) becomes too large (due to the increase of the epitaxial layer thickness during growth and thus due to the increase in strain in the structure) the dislocation will become mobile, producing in it's wake a certain length of a misfit dislocation line. Schematically, this is shown in figure 8.

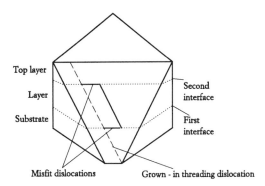

Fig. 8.    A threading dislocation glides within the epi–layer leaving in it's wake a misfit dislocation. From reference 51.

Matthews and Blakeslee defined the so-called "critical' layer thickness: the maximal thickness of the epitaxial layer which can accomodate misfit strain by elastic deformation of the lattice, i.e. without giving rise to misfit dislocations. According to mechnical equilibrium theory, this critical thickness is given by

$$h_c = \frac{b}{2\pi f}\frac{1-\nu}{1+\nu}\ln\left[\frac{h_c}{b}+1\right] \tag{13}$$

Here, b is the Burgers vector of the dislocation, $\nu$ is Poisson's ratio (the ratio of compliances) and $h_c$ is the critical layer thickness. We make note here that the value of 2 in the denominator of the first term can be 4 or 8 depending on whether the strained layer is just a single layer, is capped by a top layer with a lattice constant equal to the substrate or whether the strained layer is incorporated in a superlattice, respectively.

We stress the fact that the equilibrium theory provides a minimal value of the critical layer thickness as it does not take into account any nucleation barrier towards the formation of a misfit dislocation. The model of Matthews and Blakeslee considers only the glide of a threading dislocation, which, in the f.c.c. lattice, takes place through the {111} planes. Later models have examined also the spontaneous nucleation of locations. There are three main types of dislocations: the edge-, screw- and 60⁰ dislocations, while also stacking faults (partial dislocations) have to be considered as a possible lattice-imperfection in strained layers. The edge dislocation has the highest core energy and thus the highest activation energy for nucleation [52]. The screw dislocation has a low activation energy for nucleation but cannot relax tetragonal distortion such as will occur in a strained cubic-lattice. Thus, the 60⁰ perfect dislocation is the most probable type of misfit dislocation in diamond-type and zink-blende-type of lattices e.g. in the Si–Ge materials system as well as in the III–V and II–VI materials.

The 60⁰ perfect dislocation has a Burgers vector $\mathbf{b} = \frac{1}{2}a <110>$ and can thus be built up out of two partial Schockly dislocations according to:

$$\underset{\mathbf{b}}{\tfrac{1}{2}a[101]} \longrightarrow \underset{\mathbf{b_1}}{\tfrac{1}{6}a[21\bar{1}]} + \underset{\mathbf{b_2}}{\tfrac{1}{6}a[\bar{1}12]} \qquad\qquad (14)$$

This is schematically shown in figure 9, the projection of a (111) plane in a f.c.c. lattice.

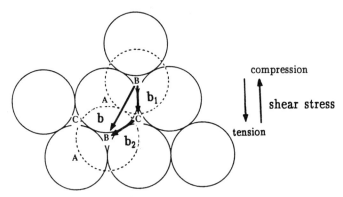

Fig. 9. Projection of a (111) plane in a f.c.c. lattice. The interrupted circles indicate the topmost layer of atoms; A, B and C the stacking sequence. The Burgers vectors are indicated as well as the direction of the shear stress. From reference 53.

Note that the $\mathbf{b_1}$ partial dislocation is parallel to the shear stress. Thus, the $\frac{1}{6}a[21\bar{1}]$ partial requires only a small shear stress to be formed and that the nucleation occurs with tensional strain ($\Delta f < 0$). With growth in compression, the $\frac{1}{6}a[\bar{1}12]$ has to be nucleated first. As this partial dislocation feels a smaller shear stress a larger amount of strain has to be applied before slip in this direction can occur. However, once $\mathbf{b_2}$ is formed, the amount of stress present in the layer is large enough to provide movement along $\mathbf{b_1}$ as well and the total displacement is according to the direction of $\mathbf{b}$, i.e. the perfect 60⁰ dislocation. Thus, in compressional strain, a relatively large force is required which results in a perfect 60⁰ misfit dislocation, while the stacking sequence of {111} planes remains ....ABCABCABC.... as required in a perfect f.c.c. lattice.

Tensional strain, however, will initially nucleate only the partial dislocation. With displacement along $\mathbf{b_1}$, the stacking sequence in {111} can become ....ABCABCBABC.... i.e. gives rise to only a partial dislocation in the form of a stacking fault, or, when the sequence ....ABCABCBACBAC.... is formed, gives rise to "twinning", i.e. crystalline ordering in a mirror image.

The dislocations which are present in crystalline structures can readily be observed by means of transmission electron microscopy (TEM). As example, a bright field TEM micrograph of a MOVPE grown GaP-GaAs$_{.33}$P$_{.67}$-GaP double heterostructure is shown in figure 10. In this sample both the GaAs$_{.33}$P$_{.67}$ layer as well as the GaP top layer were grown to a thickness of 530 Å. These thicknesses are well above the critical layer thickness according to equation (13) as the mismatch between GaAs$_{.33}$P$_{.67}$ and GaP amounts to 1.2 %, which according to (13) gives a value of h$_c$ of approximately 100 Å. The sample is viewed with the (001) interfaces viewed at a slight angle from parallel.

A planar view TEM image of a compressively grown interface is shown in figure 11.

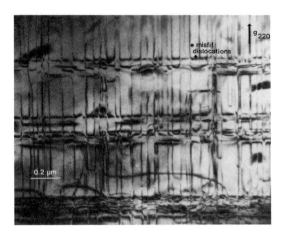

Fig. 10.   Bright field micrograph of a double hetero– structure displaying partial– and perfect misfit dislocations in the tensile (top) interface, respectively the compressive (bottom) interface. From reference 53.

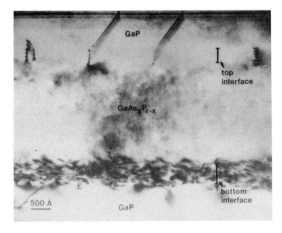

Fig. 11.   TEM planar view image of a GaAs$_{.33}$P$_{.67}$ on GaP bottom interface, with the GaAs$_{.33}$P$_{.67}$ layer grown above the critical layer thickness. From reference 54.

The pattern at the compressively grown interface reveals a rectangular grid of misfit dislocations which show a certain amount of curvature at the intersections. This curvature must be attributed to interaction of the dislocations lying in either the

$\{111^A\}$ and $\{111^B\}$ planes intersecting the (001) surface along the <110> directions. A more detailed investigation of dislocation nucleation in such GaP-GaAs$_x$P$_{1-x}$-GaP double heterostructures was carried out in [51] making use of low–temperature cathodo–luminescence (LTCL). A large number of samples, with varying As content and varying thickness were investigated. It was found that misfit dislocations were formed when the strained layer thicknesses was above the value as calculated from equation (13). Furthermore, it was found that the dislocations initially nucleated along one <110> direction giving rise to a parallel array of dislocation lines. Only at values three to five times above h$_c$ a square (rectangular) grid of misfit dislocation lines, running in both <110> directions, is found. An overview of these LTCL results is shown in fig. 12.

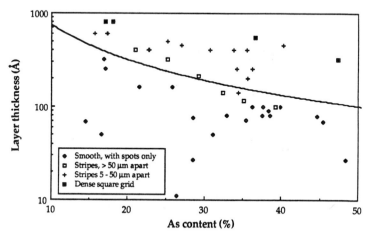

Fig. 12.    Summary of LTCL investigations on GaP-GaAs$_x$P$_{1-x}$– GaP structures. The curve represents the value of h$_c$ as calculated from (13). From reference 51.

From this plot, two important conclusions can be drawn.
First of all: the formation of misfit dislocations occurs at values very close to those predicted by equilibrium theory. Thus, one can conclude that the nucleation barrier to formation of misfit dislocations is negligible in this materials system. In turn, this implies that formation of misfit dislocations takes place by displacement of threading dislocations. Next, in [51] it has been established that the misfit dislocations are formed preferentially in one of the <110> directions. This is in agreement with previous findings [55] and can be understood by the difference between the $\{111^A\}$ and $\{111^B\}$ planes in a III–V lattice: The $\{111^A\}$ planes contain only group III atoms, the $\{111^B\}$ planes only group V atoms. The interatomic binding (and thus the barrier to slip) is smaller in the $\{111^A\}$ planes than in the $\{111^B\}$ planes so that perfect misfit dislocations will glide preferentially through the "A–set". The negligable activation energy for nucleation and the easy slip through $\{111^A\}$ result in formation of a parallel array of dislocations at an epilayer thickness set by the equilibrium theory. However, in the Si–Ge materials system, the difference between $\{111\}$ planes does not exist. Also, the threading dislocation densities of silicon substrates are much lower than in substrates of the III–V materials. Thus, in Ge$_x$Si$_{1-x}$ strained layers, misfit dislocations have to be nucleated spontaneously and not necessarily at threading dislocations. A first evaluation of the critical layer thickness in Ge$_x$Si$_{1-x}$ stained layers appeared in [56]. A new equation for the critical layer thickness as a function of mismatch was derived based on the assumption that dislocations will nucleate spontaneously when the areal strain density of the epitaxial layer exceeds the self-energy of an isolated dislocation. In [56] the nucleation of the screw dislocation as isolated dislocation was proposed resulting in an expression for the critical layer thickness as follows:

$$h_c = \frac{b^2}{16\pi \sqrt{2} \, f^2} \frac{1-\nu}{1+\nu} \frac{1}{\bar{a}} \ln \frac{h_c}{b} \tag{15}$$

in which $\bar{a}$ is the mean lattice constant and the other symbols have already been defined. Comments on this paper have been given in [57]; here we only make note of the fact that the screw dislocation cannot relieve tetragonal strain and that the $60^0$ perfect dislocation is the most favoured type for accommodation of compressive strain. Maree et al. [58] have also considered the spontaneous nucleation of dislocations in order to relieve strain and derived equations for the critical layer thickness. In this paper also distinction is made between compressive and tensile strain. The compressive strain gives rise to formation of perfect $60^0$ dislocations via the nucleation of half–loops while the tensional strain gives rise to stacking faults. The mechanism of formation and propagation of half-loop dislocations is shown schematically in figure 13.

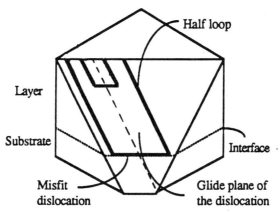

Fig. 13.    Schematic illustration of formation and propagation of half-loops. The dotted line indicates a threading dislocation (not necessary).

Nucleation of the dislocation is from a surface source, this need not be at a threading dislocation outcrop. Propagation of the half-loop is through the {111} glide planes. The resulting misfit dislocation is of the $60^0$ type. In [59], the results are described of investigations in the $Ge_xSi_{1-x}$ materials system with respect to the threading dislocation density of the substrate. These investigation concern the region between nucleation of misfit dislocations at a threading dislocation with zero activation energy, and spontaneous nucleation of half-loops, from surface sources, for compressively strained layers. It is shown that a metastable region exists, where no dislocations exist although the epitaxial layer is above the equilibrium value where it is energetically favourable to relieve mismatch by dislocations. Furthermore, it was observed that mismatch was relieved "patchwise", with high dislocation density and large strain relief in regions around existing threading dislocations. Also, it was demonstrated that several half loops could be generated around one single threading dislocation.

## FINAL REMARKS

Several equations have been proposed for the estimation of the critical layer thickness of an epitaxial layer grown with a certain amount of mismatch to the substrate. The mechanical equilibrium theory, as formulated by Matthews and Blakeslee sets the lower limit of $h_c$. The spontaneous nucleation theories give higher values for $h_c$ although one must realize that strained layers grown to a thickness above the thermodynamic equilibrium value are in principle only metastable layers. Strained layers in the III-V systems ($GaAs_xP_{1-x}$ on GaP [51] and $In_xGa_{1-x}As$ on GaAs [60]) have not been grown as metastable structures and misfit dislocations are formed at the thermodynamic equilibrium thickness. This can be understood by the fact that the III-V structures are grown on substrates with high dislocation density ($\sim 10^5$ cm$^{-2}$) and nucleation of the misfit dislocation involves a negligible activation energy. Once formed, the misfit dislocation propagates to the epilayer-substrate interface via half loops which glide through the {111} planes. The difference between the {111$^\text{A}$} and {111$^\text{B}$} planes in the III-V lattice result in, initially, slip in only one set of these planes. For release of strain in a compressively grown structure (f > 0) the majority of misfit dislocations are of the perfect, 60$^0$ type, with Burgers vector $\pm\frac{1}{2}$a [101].

Ge$_x$Si$_{1-x}$-Si structures can be grown (in compression) on substrates with a low density of threading dislocations. In this system, nucleation is either at a threading dislocation or spontaneously, from a surface source. The spontaneous nucleation may be retarded due to the fact that a certain activation energy for dislocation nucleation has to be built up by excessive strain. Without threading dislocations present, the most likely mechanism of dislocation formation and propagation is via half-loops [58], resulting in perfect 60$^0$ dislocations at the epilayer-substrate interface for compressively grown structures. Recently, also dislocation propagation in V-shape has been discussed [61]. Here, it is proposed that the most-likely mechanism of strain relief via spontaneous nucleation is by a combination of the perfect 60$^0$ dislocation with a screw dislocation. The screw dislocation alone cannot relieve the tetragonal strain encountered with f.c.c. lattices.

De 60$^0$ dislocation can be formed by two partial dislocations with Burgers vector $\frac{1}{6}$a<112>. Of these two, the 90$^0$ partial is parallel to the strain field and experiences the largest force i.e. nucleates more easily. With tensional strain, stacking faults can be formed relatively easy as a result of the occurrence of only the 90$^0$ partial dislocation. With compressive strain the 30$^0$ partial is nucleated first and is followed immediately by the 90$^0$ partial, so that the 60$^0$ perfect dislocation is formed. As a relatively high shear stress is required for the 30$^0$ partial, the critical layer thickness for layers grown in compression is slightly above the value for layers grown in compression. Thus with the various different models for the nucleation of misfit dislocations a number of different relations are found between the critical layer thickness versus lattice mismatch. Four theoretical curves are shown in figure 14.

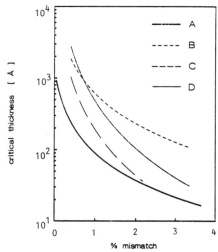

Fig. 14.  The critical thickness versus lattice mismatch. Curve A: equilibrium theory. Curve B: V-shaped dislocations. Curve C: compressive strain, half-loop nucleation. Curve D: tensile strain, partial dislocations c.q. stacking faults.

As can be seen from this figure, also the metastable layers have a limited thickness at a larger degree of mismatch. The most promising way to expand the range in which strained layers can be grown is by using patterned substrates [62]. When the growth area of the epitaxial layer is reduced one can significantly reduce dislocation interactions and this will increase the metastability region for strained layers. As a final remark, we must add that strain has a profound effect on the efficiencies of element incorporation in a.o. ternary III-V layers. Thus incorporation anomalies have been reported on for MBE grown $Al_{1-x}In_xAs$ [63] and MOVPE grown $GaAs_xP_{1-x}$ [54].

## SUMMARY

In this paper we have discussed the process of metal organic vapour phase epitaxy (MOVPE) with an emphasis on the fundamental aspects of the technique. In the introduction, the materials which can be grown by MOVPE are listed and the basic principles of the method are explained. A schematic diagram of the equipment needed for growth of GaAs/(Al,Ga)As structures is given.

In the second section the various reactor designs are discussed in relation to the two types of fluid motion, namely forced and free convection. It has been pointed out that operating the VPE reactor at reduced pressure will minimize the importance of the carrier gas flow behaviour. Thus, in vacuum chemical epitaxy and in metal organic molecular beam epitaxy, the problems related to mass transfer by a carrier gas are completely eliminated.

Next, an inventory is given of the metal organic components which may be used for growth and/or doping. Also, the decomposition mechanism of trimethylgallium is discussed and the experimental data on the growth mechanism of gallium arsenide are reviewed. It is clear that the decomposition mechanism of the metal organic molecules involves a series of steps in a chain reaction, and that free radicals are the intermediate products in the sequential dealkylation of the starting compounds. With "typical" growth conditions (substrate temperature 700$^0$C, partial pressure of TMGa $\leq 10^{-4}$ atm., $P_{AsH3} \sim 2 \cdot 10^{-3}$ atm.) the most simple description is that the growth rate is limited by the diffusion of the gallium component through the boundary layer adjacent to the susceptor surface. During diffusion, TMGa partially decomposes to monomethylgallium and these species eventually arrive on a surface consisting of adsorbed AsH. Removal of the final methyl group takes place via the hydrogen from the adsorbed AsH and consequetively, crystalline GaAs is formed. The growth at high temperatures results in low epitaxial growth rates due to desorption of gallium from the surface. With growth at low temperatures, desorption of reaction products and/or hydrogen is the rate-limiting step. Thus, the mechanism of GaAs growth by MOVPE, using trimethylgallium and arsine, can be reasonably well given: the gas phase reaction has been accounted for and it has been shown that the carbon contamination in epitaxial GaAs layers is typically below the level of $10^{15}$ cm$^{-3}$ due to a reaction of the final methyl group of TMGa with atmomic hydrogen, bound on the surface via the AsH species. The final methyl group can also be driven off the surface by arsenic species, but a residual carbon level of $10^{20}$ cm$^{-3}$ remains in GaAs films grown with TMGa and As$_4$.

The situation regarding AlAs growth is less clear. At present, no reliable kinetic data are available on the pyrolysis of trimethylaluminium. At present, no organo-aluminium precursor compatible with TMGa is available for the growth of $Al_xGa_{1-x}As$ layers with low impurity concentration. It seems worthwhile to investigate further alternative compounds, as discussed in [64] and [65].

## FUTURE DIRECTIONS

Clearly, the process of MOVPE offers a great potential for applications and research. The flexibility of the technique to grow a complete range of III-V and II-VI materials has been noted. Cross- disciplinary interaction with the field of fluid dynamics leads to optimization of reactor designs, whereas interaction with the field of metal organic chemistry leads to the development of new starting compounds for growth and doping .

The use of metal organic compounds in an MBE system yields extremely good results, notably in improvement of the surface morphology due to the elimination of the oval defects and in the very reproducible growth of the phosphorous containing materials [16].

The process of MOVPE is now also applied in growth via the so–called atomic layer epitaxy. To this end, novel reactors have been designed [66] to sequentially deposit only one atomic layer of one component per cycle. By atomic layer epitaxy, extremely good thickness control can be achieved which will be most important for the growth of superlattice structures [67].

Furthermore, we have discussed the possibilities and problems related to the growth of strained layer structures. Such materials give the scientist more flexibility, a.o. to vary the band–gap of the semiconductor independently of the lattice constant.

Finally, understanding of the fundamental processes in MOVPE will greatly benefit from experiments with in situ monitoring of the gas phase composition and/or the epitaxial surface structure during growth. For a determination of the chemical species in the gas phase, real time diagnostics with i.r. laser diode spectroscopy has been shown to be a valuable tool [38]. Also, Raman scattering spectrometry can be employed to investigate both the temperature distribution as well as the type and concentration of chemical species present in the gas phase [39]. RHEED investigations, carried out during CBE growth give new insight in the decomposition– and desorption kinetics of the metal organic molecules [68]. More recently, the technique of reflective–difference spectroscopy has been used to obtain chemical– and structural information on growing surfaces [69], [70].

Parts of this paper have already been published, namely refs. 4, 44 and 45. These parts are reproduced with kind permission from the publishers, Butterworth–Heineman Ltd.

## REFERENCES

[1]  H.M. Manasevit, J. Electrochem. Soc. 116 (1969) 1725
[2]  R.D. Dupuis and P.D. Dapkus, Appl. Phys. Lett. 31 (1977) 466
[3]  M. Razeghi, B. de Cremoux and J.P. Duchemin, J. Cryst. Growth 68 (1984) 389
[4]  M.R. Leys, Chemtronics 2 (1987) 155
[5]  S.J. Bass, J. Cryst. Growth 31 (1975) 72
[6]  J.P. Duchemin, M.Bonnet and F. Koelsch, J. Electrochem. Soc. 126 (1979) 1134
[7]  P.M.Frijlink, J. Cryst. Growth 93 (1988) 207
[8]  J. van Suchtelen, J.E.M.Hogenkamp, W.G.J.M. van Sark and L.J.Giling, J. Cryst. Growth 93 (1988) 201
[9]  F.C. Eversteijn, P.J. Severin, C.H.J. van den Brekel and H.L. Peek, J. Electrochem. Soc. 117 (1970) 925
[10] K.F. Jensen, J. Cryst. Growth 98 (1989) 148
[11] R. Jet Field, J. Cryst. Growth 97 (1989) 739
[12] J.H. Van der Ven, G.J.M. Rutten, M.J. Raymakers and L.J. Giling, J. Cryst. Growth 79 (1986) 352
[13] C. van Opdorp and M.R. Leys, J. Cryst. Growth 84 (1987) 271
[14] C.A. Wang, S. Patnaik, J.W. Caunt and R.A. Brown, J. Cryst. Growth 93 (1988) 228
[15] L.M. Fraas, J. Electron. Mater. 15 (1986) 175
[16] W.T. Tsang, J. Cryst. Growth 98 (1989) 226
[17] Gaskill, D.K. et al., J. Cryst. Growth 77 (1986) 418
[18] Ludowise, M.J. and Cooper, C.B., SPIE 1982, 323, 117
[19] A. Brauers, O. Kayser, R. Kall, H. Heinecke, P. Balk and H. Hofman, J. Cryst. Growth 93 (1988) 7
[20] G.B. Stringfellow, J. Cryst. Growth, proceedings ICMOVPE 5, (1990)
[21] Bass, S.J., J. Cryst. Growth 47 (1969) 613
[22] H.M. Manasevit and A.C. Thorsen, J. Electrochem. Soc. 119 (1972) 99
[23] Glew, R.W., J. Cryst. Growth 77 (1984) 44
[24] C.H. Chen, et al., J. Cryst. Growth 77 (1986) 11
[25] Glew, R.W., J. de Physique 43 (1982) 281
[26] T.F. Kuech, E. Veuhoff and B.S. Meyerson, J. Cryst. Growth 68 (1984) 48
[27] A.P. Roth, R. Yakimova and V.S. Sundaram, J. Cryst. Growth 68 (1984) 65
[28] J.D. Parsons and F.G. Krajenbrink, J. Cryst. Growth 68 (1984) 60
[29] D.W. Kisker, J. Cryst. Growth 100 (1990) 126

[30] J.A. Long, V.G. Riggs, A.T. Macrander and W.D. Johnston, J. Cryst. Growth 77 (1986) 42
[31] M. Akiyama, Y. Kawarada and K. Kaminishi, J. Cryst. Growth 68 (1984) 39
[32] V. Aebi, C.B. Cooper, R.L. Moon and R.R. Saxena, J. Cryst. Growth 55 (1981) 517
[33] J. Weber et al, J. Cryst. Growth 100 (1990) 467
[34] D.J. Schlyer and A.J. Ring, J Electrochem Soc. 124 (1977) 569
[35] M.R. Leys and H. Veenvliet, J. Cryst. Growth 55 (1981) 145
[36] D.H. Reep and S.K. Gandhi, J Electrochem Soc. 130 (1983) 675
[37] M. Tirtowidjojo and R. Pollard, J. Cryst. Growth 93 (1988) 108
[38] D.K. Gaskill, V. Kolubajev, N. Bottka, R.S. Sillmon and J.E. Butler, J. Cryst. Growth 93 (1988) 127
[39] R. Luckerath, P. Tommack, A. Hertling, H.J. Koss, P. Balk, K.F. Jensen and W. Richter, J. Cryst. Growth 93 (1988) 151
[40] K. Saito, E. Tokumitso, T. Akatsuka, M. Miyauchi, T. Yamada, M. Konagai and K. Takahashi, J. Appl. Phys. 64 (1988) 3975
[41] R. Bhat, J. Electron. Mater. 14 (1985) 433
[42] T.F. Kuech, E. Veuhoff, T.S. Kuan, V. Deline and P. Potemski, J. Cryst. Growth 77 (1986) 257
[43] N. Kobayashi and T. Makimoto, Jpn. J. Appl. Phys. 10 (1985) L824
[44] M.R. Leys, Chemtronics 3 (1988) 179
[45] M.R. Leys, Chemtronics 4 (1989) 31
[46] L.M. Yeddanapalli and C.C. Schubert, J. Chem. Phys. 14 (1946) 1
[47] A.J. Quimet, Dissertation University of Connecticut, USA, 1962
[48] G.C. Osbourne in Semiconductors and Semimetals, Academic Press, New York (1987) 459
[49] M.E. Pistol, M.R. Leys, L. Samuelson, Phys. Rev. B37 (1988) 4664.
[50] J.W. Matthews in: Epitaxial Growth, Part B, Ed. J.W. Matthews, Academic Press, New York (1975)
[51] A. Gustafsson, Masters Thesis, University of Lund, Sweden (1990)
[52] D. Hulland and D.J. Bacon, Introduction to dislocations, 3rd ed., Pergamon Press, Oxford (1984)
[53] J. Petruzello, M.R. Leys, Appl. Phys. Lett. 53 (1988) 2414.
[54] M.R. Leys, H. Titze, L. Samuelson, J. Petruzello, J. Cryst. Growth 93 (1988) 504.
[55] G.H. Olsen, M.S. Abrahams and J.J. Zamerowski, J. Electrochem. Soc. 121 (1974) 650.
[56] R. People, J.C. Bean, Appl. Phys. Lett 47 (1985) 322
[57] B.W. Dodson, P.A. Taylor, Appl. Phys. Lett. 49 (1986) 642
[58] T.M.J. Maree, J.C. Barbour, J.F. van der Veen, K.L. Kavanagh, C.W.R. Bulle-Lieuwma and M.P.A. Viegers, J. Appl. Phys. 62 (1987) 4413
[59] C.G. Tuppen, C.J. Gibbings and M. Hockly, Journ. Cryst. Growth 94 (1989) 392
[60] T.G. Andersson, Z.G. Chen, V.D. Kulakovskii, A. Uddin, J.T. Vallin. Appl. Phys. Lett. 51 (1987) 752
[61] Y. Fukuda, J. Cryst. Growth 100 (1990)
[62] E.A. Fitzgerald, J. Vac. Sci. Technol. B7 (1989) 782
[63] F. Turco and J. Massies, Appl. Phys. Lett. 51 (1987) 1989
[64] A.C. Jones, P.J. Wright, P.E. Oliver, B. Cockayne and J.S. Roberts, J. Cryst. Growth 100 (1990) 395
[65] L. Pohl, M. Hostalek, H. Luth, A. Brauers and F. Scholz, J. Cryst. Growth, proceedings IC MOVPE V, to be published
[66] S.M. Bedair et al., Appl. Phys. Lett. 47 (1985) 51
[67] M. Ozeki, N. Ohtsuka, Y. Sakuma and K. KodamaJ. Cryst. Growth, proceedings IC MOVPE V, to be published
[68] A. Robertson, T.H. Chiu, W.T. Tsang and J.E. Cunningham, J. Appl. Phys. 64 (1988) 877
[69] D.E. Aspnes, R. Bhat, E. Colas, V.G. Keramidas, M.A. Koza and A.A. Studna, J. Vac. Sci. Technol. A7 (1989) 711
[70] J. Jonsson, K. Deppert, S. Jeppesen, G. Paulsson, L. Samuelson and P. Schmidt, Appl. Phys. Lett. 56 (1990) 1

## SUBMICRON PATTERNING TECHNIQUES FOR INTEGRATED CIRCUITS

W. Beinvogl and A. Gutmann

Siemens AG, Semiconductor Group
Otto-Hahn-Ring 6, D-8000 Munchen 83
Germany

### INTRODUCTION

The development of integrated circuits with ever increasing density has pushed the requirements on fine patterning technology into the submicron region during the late 80's and device design rules will cross the 0.5 µm barrier on a production level towards the middle of this decade. Optical lithography is the by far prevailing method presently, x-ray lithography has entered the pilot line evaluation phase. Whether or at what time optical lithography might lose its dominating role in IC mass production remains an open question to be answered by the upcoming technological improvements and the economic performances of the lithographic techniques competing with each other. The topics discussed in this paper are limited to optical lithography which will maintain it's position as the industrial workhorse at least for several years. After a discussion of the lithography requirements in the next section, a discussion of resist materials for the relevant wavelength regions followed by examples for enhanced lithographic processes will be given.

For transferring the lithographically generated patterns into underlying materials dry etching became the standard method during the 80's. This field is characterized by an increasing diversity both with respect to technical methods and involved chemistries due to the large number of different materials to be etched. In addition to the wellknown patterning of various layers into fine patterns, deep trench/groove etching into the semiconductor substrate as well as blanket planarizing etchback became important applications. These two items will be addressed in the section on dry etching after a short view over requirements and methods in the dry etching field.

### LITHOGRAPHY REQUIREMENTS FOR A MANUFACTURABLE PROCESS

Forwarding the required resolution capability is the prerequisite for the employment of any lithographic technique, but only one major issue among a long list of multifold requirements on a production process as shown in table 1 (with no claim to completeness). The theoretical limit of resolution in projection printing is generally described by the Rayleigh criterion $d_{min} = \frac{1}{2} \cdot \frac{\lambda}{NA}$ . This formula indicates already the twofold strategy in optical lithography to meet the demands of ever decreasing ground rules, namely rising numerical apertures (NA) and/or the shift to shorter exposure wavelengths. The improvement of exposure equiment is going hand in hand with a continuous optimization of resist materials and resist processing.

Table 1. Requirements on a lithograpic production process

**resolution**
(provided by lens and resist system)

**requirements on optics**
- field size
- image quality (minimization of lens errors)
- stability of light source
- sufficient intensity

**overlay accuray**
- minimization of mechanical misalignments
- small lens distortion
- small reticle rotation errors
- precision of mask
- good quality of alignment marks

**machine-to-machine matching of
exposure/alignment tools**

**requirements on resist system**
- good adhesion
- thermal stability
- etch resistance
- strippability
- low metal ion content

**accuracy of parameter control equipment**

**wide process latitudes, e. g.**
- exposure dose latitude
- defocus latitude
- latitude with respect to time intervals between
  process steps

**retention of ground rules across wafer
topography**

**retention of ground rules over substrates
of high reflectivity**

**uniformity/repeatability of processes**

**stability/homogeneity of materials**

**low particle density**

**economic considerations**
- costs of investment, maintenance, materials, labor
- minimization of process complexity
- high throughput
- high uptime
- high yield

R&D data on these subjects often refer to patterning on flat substrates in the centre of the imaging field. In production critical dimension (CD) variations have to be kept within defined limits across the whole image field, even at high topography or enhanced substrate reflectivity.

The practical resolution limit achievable under manufacturing conditions is generally estimated using the formula $d_{prod} = k_1 \cdot \frac{\lambda}{NA}$. The factor $k_1$ depends on the lithographic process.

Table 2. Optical lithography of presence and near future: Resolution as a function of $\lambda$, NA and $k_1$.

| | | g-line | i-line | deep UV | |
|---|---|---|---|---|---|
| | **wave-length** | $\lambda = 436$ nm | $\lambda = 365$ nm | $\lambda = 248$ nm(KrF) | 240-265 nm |
| | **exposure equipm.** | **stepper** | **stepper** | **stepper** | **step-and-scan** |
| late 80's | NA | 0.42 - 0.55 | 0.35 - 0.48 | 0.37 - 0.42 | 0.35 |
| | resolution | | | | |
| | 0.5 $\lambda$/NA | 0.52 - 0.40 | 0.52 - 0.38 | 0.34 - 0.30 | 0.36* |
| | 0.8 $\lambda$/NA | 0.83 - 0.63 | 0.83 - 0.61 | 0.56 - 0.47 | 0.58 |
| estimate for early 90's | NA | 0.60 - 0.65 | 0.50 - 0.55 | 0.45 - 0.50 | 0.40 |
| | resolution | | | | |
| | 0.5 $\lambda$/NA | 0.36 - 0.34 | 0.37 - 0.33 | 0.28 - 0.25 | 0.32 |
| | 0.8 $\lambda$/NA | 0.58 - 0.54 | 0.58 - 0.53 | 0.44 - 0.40 | 0.51 |
| | 0.6 $\lambda$/NA | 0.44 - 0.40 | 0.44 - 0.40 | 0.33 - 0.30 | 0.38 |

*.....approximation of resolution values calculated for $\lambda = 253$ nm

Table 3. Depth of focus budget for patterning 0.6 μm line/space structures

| contribution<br>positiv ( + ) / negativ (-) | | high contrast g-line resist (1.5 μm),<br>NA = 0.55, $\varnothing_{lens}$ = 28.3 mm |
|---|---|---|
| lens/resist system<br>(centre of imaging field,<br>flat substrate) | ( + ) | 1.60 μm[*], |
| optical thickness of resist | ( - ) | |
| short term focus repeatability | ( - ) | |
| field curvature | ( - ) | 0.40 μm[1] |
| astigmatism | ( - ) | 0.20 μm |
| wafer flatness<br>(with die-by-die levelling) | ( - ) | 0.30 μm |
| long term focus repeatability/<br>focus setting accuracy | ( - ) | 0.30 μm |
| allowance for topography,<br>wafer warpage, focus accuracy<br>deterioriation | ( - ) | 0.40 μm |

[*]...average value of 4 investigated high contrast resist systems
chosen criteria: a) CD tolerance = ± 0.06 μm
b) tolerance for reduction of resist height = 150 nm

For single level techniques with conventional positive mode resists a $k_1 \sim 0.8$ has been found empirically, while $k_1 = 0.6$ is appropriate for enhanced resist schemes.

Theoretical and practical resolution limits are listed in tab. 2 in consideration of the exposure wavelengths and exposure tools applied at present or predicted for near future production processes. Projecting these data on requirements for DRAM generations, it can be said, that the g- and i-line exposure wavelengths will dominate 16M DRAM manufacturing. A major impact of deep UV technology can be expected for the 64M generation.

Wide process latitudes are a key issue for manufacturibility. Finding sufficient defocus latitude has become a major problem of optical lithography in view of the observed trends towards shorter exposure wavelength and higher NA, as estimated from the Rayleigh depth of focus criterion $DOF = \frac{\lambda}{NA^2}$. The different contributions to a DOF budget depending on parameters of the exposure/levelling tool, the resist system and wafer processing are listed in table 3 for the case of high contrast g-line resists and NA 0.55. Enhanced lithographic methods such as multilayer or surface imaging techniques have been developed as back-up solutions, in case single layer processing (being obviously the favourite from the economic point of view) should not meet any longer DOF requirements. Optimization of device design and processing and the introduction of new planarization techniques such as chemical mechanical polishing (2) represent other strategies to attenuate the DOF problem. Another recent concept proposes the application of the socalled "focus latitude enhancement exposure" (FLEX), i.e. the sequential exposure in several different focal planes. Significant increases in DOF, especially for the patterning of hole structures, are claimed in the literature (3).

Printing small features across topography with usable process latitude is one major task of a device production process, but in addition the geometries have to be placed precisely in positions predetermined by patterns of preceeding process levels. As a rule of thumb it can be said that overlay accuracy (3σ value) should be about one third of the minimum feature size. A general reduction of the alignment error by 0.05 μm would allow a shrinkage of device area by 20 - 30 %.Thus improvement of overlay accuracy has to come along with smaller design rules and will be achieved by simultaneous improvements of stage positioning, reticle alignment and alignment mark detection, furthermore by reductions of reticle errors and lens distortion.

## LITHOGRAPHIC TECHNIQUES

The present level of performance of resist materials and modifications in processing related to single layer applications are discussed shortly in the first part of this section. In the case of deep UV resists more attention will be given to varying strategies of resist design. A vast number of enhanced single or multilayer processes of varying complexity have been developed in the meantime such as image reversal schemes, contrast enhancement layers (CEL) or bi- or trilevel techniques, just to mention a few. Within the chapter on enhanced lithographic processes we will limit ourselves to the discussion of three techniques offering high potential for improvement without excessive process complexity, having therefore reasonable chances to find wider acceptance in the production environment.

### Single Layer Processes

g- and i-line Resists: Extensive R&D work (e.g. 4, 5) concerning the complex interactions between the different components of the resist/developer system have laid the foundation for a continuous improvement of near UV photoresist materials surmounting many of the previously perceived barriers of resolution. The fine-tuning in resist design included, for example, the optimization of the molecular weight distribution, of the isomeric structure of cresols, or modifications of photoactive compounds (being of the diazonaphthoquinone type for near UV resists). Providing a proper balance between strong intramolecular and intermolecular hydrogen bonding was found to furnish good dissolution selectivity as well as acceptable dissolution rate in developers. Intensified synthesis work in the i-line sector has practically closed the former performance gap between g- and i-line resist materials. In the meantime the majority of recently introduced g- and i-line products exhibit resolution at or within 10 % to the theoretical resolution limit. Resolution close to 0.4 μm could be achieved, for example, with a high contrast g-line resist when exposed with a stepper of NA = 0.55, as shown in fig. 1. Resolution down to 0.32 μm was demonstrated recently for an i-line resist when applying an experimental i-line stepper with an extremely high NA of 0.65 (6).

Employing a post-exposure bake (PEB) has become a widely accepted procedure when patterning submicron features. This bake step following exposure (at temperatures 10 - 20° higher than the prebake temperature) reduces the standing wave effect via thermally induced diffusion of the photoactive compound, thus avoiding the appearance of rippled resist profiles or even protruding lateral ridges along the foot of photoresist sidewalls. The merits of PEB are shown in particular when patterning contact holes across topography leading to improved CD control and wider process latitudes.

Incorporation of dyes in resist materials remains the least complex strategy to attenuate the problem of reflective notching which, however, may fail in cases of extremely high substrate reflectivity. Dye selection occurs based on acceptable solubility characteristics, appropriate absorptivity and sufficiently

resist:    AZ6215, 1.5 µm
exposure: g-line stepper, NA = 0.55

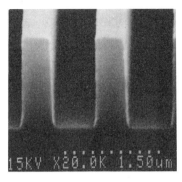

0.60 µm L/S

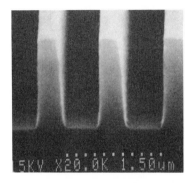

0.50 µm L/S

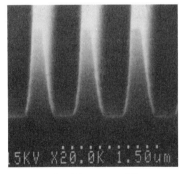

0.40 µm L/S

Fig. 1.    Patterning of line/space structures (L/S) using a g-line high contrast resist

high sublimation temperature. Negative effects of dye addition are a decrease in sidewall angle, reduced resolution and an increase in exposure dose, although all of these adverse side-effects have become less pronounced due to enhanced resist optimization.

During device production resists have to maintain profile integrity while undergoing heat-producing processes like ion implantation, RIE etching or metal deposition. With novolak-based near UV resists it is generally not possible to obtain exceptional resolution capability combined with very high thermal stability. The majority of products show a hardbake deformation temperature in the range of 120 - 140 °C (for pattern dimensions $\leq$ 5 μm). This is sufficient for most cases. If required, thermal stability can be raised to 160 - 180 °C by application of resist hardening methods like deep UV irradiation (e. g. 7) or plasma treatment (PRIST) (8).

Deep UV Resists: The major part of deep UV resist design performed during the last years focussed on the development of materials suitable for the 240 - 270 nm wave-length region. Commercial KrF (248 nm) excimer lasers have come to the market in the meantime. In addition Perkin-Elmer Company has introduced its Microscan step-and-scan catadioptric 4:1 reduction system with NA = 0.35 employing Hg Xe lamp illumination at 240 - 265 nm. Commercial near UV sensitive positive mode resists deliver sloped sidewall profiles and low sensitivity when exposed in the above mentioned deep UV range. This results from the low absorbance of diazonaphthoquinone photoactive compounds and from the high unbleachable absorbance of both photoproducts and typical near UV adapted novolak polymers. The task of developing production-compatible high resolution deep UV resists has brought a new challenge to resist designers and they answered with a wide palette of strategies. (Detailed presentations of the chemistry involved and extensive lists of references can be found in recent review articles about deep UV lithography (9, 10)).

Novolaks based on pure p-cresol have been employed offering higher deep UV transparency than common o-, m-, p-cresol mixtures. Other choices for suitable resin components have been poly-(4-vinylphenol) or derivates thereof, copolymers of styrene with maleimide or maleic acid anhydride, or silicon containing polymers. Considerable effort has been directed into the design of new highly absorbing and bleachable photoactive compounds like aliphatic or cyclic 2-diazo-1,3-diacyl or 2-diazo-1,3-dicarbonyl chromophores. However, alteration of solubility can be also achieved by means other than diazo photochemistry. Recent synthesis work has focussed on imaging systems operating via chemical amplification mechanisms. Chemical amplification involves the initial photogeneration of a catalyst which catalyses a subsequent reaction. In the case of deep UV resists the secondary reaction induces a change of dissolution properties within exposed areas. Although the quantum yield of the primary imaging process may be relatively low, the effective radiation sensitivity will increase drastically once a threshold concentration of photo-chemically generated catalyst has accumulated in order to drive the secondary reaction to completion. Fast photospeeds requiring exposure doses below 10mJ/cm$^2$ have been reported for chemically amplified deep UV resists.

With the majority of the deep UV resist systems designed so far the catalytic reaction is initiated by a photogenerated acid. Onium salts (e. g. $Ph_3 S^+ X^-$ and $Ph_3 J^+ X^-$, where $X^-$ is $AsF_6^-$, $SbF_6^-$ or $CF_3 SO_3^-$) or nitrobenzylesters of tosic acid have been used for the photogeneration of the strong acid catalyst required. Positive tone resists may operate, for example, via an acid-catalysed thermal deprotection of hydrophobic pendant groups (e.g. the t-butoxycarbonyl functionality = t-BOC) as shown schematically in fig. 2 (11). Alkaline developable negative tone resists have been synthesized too whereby crosslinking within the resin is accomplished via acid-catalysed electrophilic aromatic substitution or 0-alkylation reactions in the presence of a carbonium ion precursor (12, 13).

**Scheme A:**

**1)   photoacid generation**

$$Ar_3 S^+ X^- \xrightarrow[RH]{hv} Ar_2 S + Ar^· + R^· + H^+ X^-$$

**2)   deprotection of polymer component**

$$-(CH-CH_2)_{\overline{n}} \xrightarrow[heat]{H^+X^-} -(CH-CH_2)_{\overline{n}}$$

$$+ CO_2 + CH_2 = C \Big\langle \begin{smallmatrix} CH_3 \\ CH_3 \end{smallmatrix}$$

O CO₂t-BU          OH

in soluble in       soluble in
aqueous alkali      aqueous alkali

**Scheme B:**

novolak        $CO_2t$-Bu        novolak        $CO_2t$-Bu

$\xrightarrow{hv}$

$Ar_3S^+X^-$                          HX
in soluble in
aqueous alkali

novolak        $CO_2H$

$\xrightarrow{heat}$

HX
soluble in
Aqueous alkali

Fig. 2. Examples for chemical amplification applications for deep UV resist design.

resist:       PVP/SUCCESS (triflate)*), 1.0 μm
exposure:   deep UV stepper, NA = 0.37

 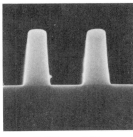 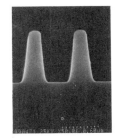

0.60 μm L/S            0.50 μm L/S            0.40 μm L/S

*)...   R. Schwalm*, H. Binder*, B. Dunbay** and A. Krause** (* BASF AG, Ludwigshafen, and ** Siemens AG, Munich), Polymers for Microelectronics, Conference, Tokyo, November 1989, proceedings to be published

Fig. 3. Patterning of line/space structures (L/S) using a positive tone deep UV resist

Employing resist thicknesses around 1.0 μm resolution capabilities in the 0.35 - 0.40 μm region have been reported in the literature for positive (compare fig. 3) as well as negative mode resists (13, 14, 15) using KrF excimer laser steppers of NA = 0.35 - 0.37. Thus, for the best cases deep UV resists already demonstrated resolution at or close to the theoretical limit. However, further time is required to check thoroughly production process compatibility. The influence of the time interval between exposure and post-exposure bake on CD variations, for example, needs careful attention when working with chemical amplification systems.

## Enhanced Lithographic Processes

Antireflection Layers (ARL): Antireflection layer methods, although generally not classified as such, belong to the group of bilayer techniques when viewed from the standpoint of complexity. These layers of either organic polymers with strongly absorbing dyes (16) or of inorganic dielectrics (17) are deposited on highly reflecting substrates prior to resist coating in order to reduce resist line notching either by absorption and/or by destructive interference between primary and reflective waves. The extent of reflectivity suppression via destructive interference depends on the angle of incidence as well as the thickness of the ARL as shown for one example in fig.4. Over a realistic topography local variations of these two parameters will occur. In practice adjustments of the nominal thickness of antireflective layers operating via interference may become necessary based on empirical data.The application of ARL's necessitates an increase in exposure dose by about 30-50%.
In combination with the use of high contrast resists the ARL technique furnishes generally steeper resist profiles and better process latitudes as compared to dyed resist applications, however, at the cost of higher process complexity. A practical example for ARL is shown in fig. 5.

Bilayer Techniques Using Silicon Containing Top Resists: Applying a bilevel approach a thin imageable layer (top resist) ist coated over a thick planarizing material (bottom resist), then exposed and wet-developed. The good short range (but not far-range) planarization of the bottom level improves focus latitude and renders better thickness uniformity and thus reduced CD variations of the top resist. Substrate reflections can be suppressed by bottom resist materials which are highly absorptive for the imagewise exposure wave-length. The image transfer from the top to the bottom layer is carried out either by deep UV flood exposure or by anisotropic reactive ion etching (RIE). The latter method has found increasing attention during the past years. It requires a silicon containing top resist which acts as an etch mask by forming a silicon oxide crust during RIE etching in $O_2$ plasma (see fig. 6). A variety of positive tone organosilicon top resists suitable for near or deep UV have been designed (e.g. 18, 19) containing Si rich novolaks or polyvinylphenol derivates, polysiloxanes with phenol pendant groups a. o. as resin components. Negative mode top resists like polysiloxanes have been used for deep UV applications (20). Incorporation of Si into a patterned silicon-free top resist via gas-phase silylation of functional groups has been also tested as a further variant of bilayer techniques.

Surface Imaging Techniques: With this type of lithographic process the functions of imaging and masking to dry-development in $O_2$ plasma are taken over by one resist material only. By choosing resins of appropriate functionalities selective incorporation of silicon within the surface-near region (100 - 200 nm) of either the exposed or unexposed resist areas can be achieved from the gas or liquid phase (20, 21, 22) resulting in negative or positive tone patterning. Since exposure is only required within the top region of the resist depth-of-focus demands become less stringent. Moreover the resist can be

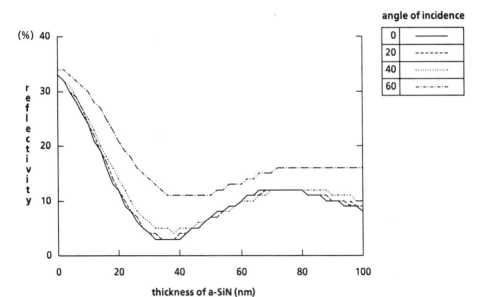

Fig. 4.    Reflectivity of a-SiN on molybdenum silicide into resist

resist system:    TSMR-V3 (1.5 μm) / NMD-W
exposure:         g-line stepper, NA = 0.55

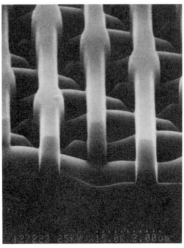

Fig. 5.    Patterning of 0.7 μm minimum geometries across molybdenum
silicide substrate using an a-SiN antireflection layer

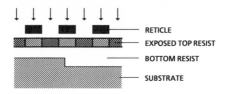

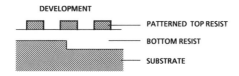

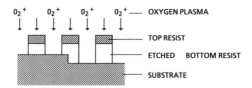

Fig. 6. Processing scheme of a bilayer technique using an organosilicon top resist.

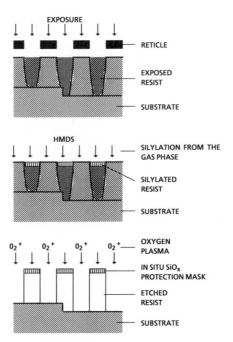

Fig. 7. Processing scheme of a surface imaging method using gas phase silylation.

heavily dyed for suppression of reflected light from the substrate. The best known among the dry-developed single-level resist schemes is the socalled DESIRE process (20) following a processing sequence as described in fig. 7. Here silylation is occurring from the gas-phase (HMDS) during a post-exposure bake step. The process has shown excellent resolution capabilities, e. g. the resolution of 0.4 µm line/space pairs when employing an i-line stepper of NA = 0.42. Extensive feasibility studies of surface imaging silylation processes for submicron manufacturing have already started.

Table 4.   Dry etching requirements for the process generations from 1.2 µm (1Mbit DRAM level) to 0.4 µm (64Mbit DRAM level)

| process generation; minimum feature size [µm] | 1.2 | 0.8 | 0.6 | 0.4 |
|---|---|---|---|---|
| wafer diameter [mm] | 125-150 | 150 | 150-200 | 200 |
| CD tolerances for lithography / etching [nm] | ± 200 | ± 150 | ± 110 | ± 100 |
| maximum allowed etch nonuniformity | ± 5% | ± 5% | ± 5% | ± 3% |
| materials to be etched | $SiO_2$ (undoped & doped), $Si_3N_4$, Poly-Si (undoped & doped), mono Si, metal silicides (e.g. Mo-, Ta-, Ti-, W-silicide), organic materials, AlSiCu, barrier layers (e.g. TiW, TiN), W, metal sandwiches | | | |
| etch selectivities required<br>Poly-Si : $SiO_2$<br>$SiO_2$ : Si<br>Al : resist | 20 : 1<br>10 : 1<br>2 : 1 | 25 : 1<br>12 : 1<br>4 : 1 | 30 : 1<br>15 : 1<br>6 : 1 | 35 : 1<br>20 : 1<br>8 : 1 |

## DRY ETCHING

The lithographically generated resist structures mainly serve as masks for subsequent etching steps. Thereby various layers are patterned usually in a selective way with respect to underlying materials.

### Requirements for Dry Etching

The requirements for these etch processes are steadily increasing (tab. 4) with shrinking minimum feature sizes (23). Each of the columns in table 4 represents a new generation in CMOS technology for high density devices like semiconductor memories. The numbers in table 4 refer to a manufacturing environment. Etch uniformities over a wafer and from wafer to wafer have to improve, a fact which is rendered more difficult by the increasing wafer diameters. CD-tolerances have to become tighter  from generation to generation even though they tend to improve more slowly then the minimum feature size.

Layer thickness of thin dielectrics, junction depths within single crystal silicon are decreasing leading to more aggressive selectivity figures without allowing any deterioration of anisotropy of course. The number of materials to be etched in a typical state-of-the art CMOS process is quite large especially in the area of metals and metal compounds. This fact results in a great diversification of process chemistries and equipment alternatives which are often tuned to the specific requirements of a particular process.

## Dry Etching Techniques

Anisotropic dry etching became common practice in the beginning eighties when features sizes approached 2 µm. Diode plasma and reactive ion etching (fig. 8) were the major equipment alternatives at that time. With reactive ion etching (RIE) at low pressure very good anisotropy can be reached generally with modest selectivity and low etchrates. With increasing demands a strive after a more selective control of the complicated physics and chemistry in the plasma lead to advanced techniques (fig. 8 bottom). In triode ion etching, which is successfully applied in deep trench etching in silicon, two independent excitations make possible the separate control of reactant density and ion bombardment energy. In magnetron RIE the additional magnetic field enhances the ionisation efficiency and thereby etchrate.

## Application Examples

In the following examples on etching problems and results are given for deep trench etching and planarisation etch, areas which are of special importance for highest density IC's.

Deep Trench Etching: Deep trench etching in silicon found several applications, the most important of which is the fabrication of capacitors for DRAM cells folded into the silicon substrate. 4M DRAM products from several manufactures make use of this approach to preserve the available charge per memory cell inspite of the decreasing cell area. Out of the many requirements for this etching process the achievement of suitable trench profiles is the most demanding one. This aspect of deep trench etching is illustrated briefly in the following. Fig. 9 shows trench profiles obtained after RIE in a $BCl_3/Cl_2$-chemistry in a batch type reactor (24). Atomic chlorine serves as reactant for the formation of volatile $SiCl_4$ in this chemistry. Depending on the pressure range and on the resulting mean free path of the ions either rough sidewalls or residual silicon pillars at the trench bottom can not be avoided. The latter result from ions specularly reflected from the sidewall thereby enhancing the etchrate along the rim compared to the center of the trench. Trench profiles like those in fig. 9 are not acceptable for a reliable formation of a capacitor.The results can be significantly improved in a $CBrF_3$-chemistry (fig. 10, (25)), where Br is the major reactant with silicon. During etching an inhibitor film of increasing thickness builds up, pinching this way the cylinder of ion reflection with increasing etch depth. The inhibitor film is easily dissolvable in HF after dry etching and consists mainly of $SiO_2$ which originates from the $SiO_2$ etch mask. Also the carbon content in the $CBrF_3$ contributes to the inhibitor film.
After inhibitor removal a perfect profile is obtained with a hemispherically shaped trench bottom. These trenches allow the formation of thin multilayer dielectrics (10-20 nm thickness) with a quality equivalent to planar capacitors as well as a complete trench filling after capacitor formation.

Planarization: One application of planarization etch is found in the buried oxide isolation scheme ("BOX") suitable for submicron dimensions (26). Shallow trenches are etched into the silicon to isolate the active regions from each other.

| Method | Downstream etch | Diode plasma etch | Reactive ion etch |
|---|---|---|---|
| Principle | 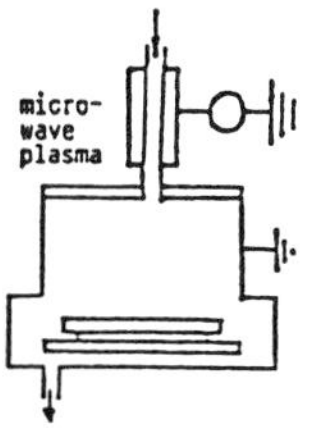 |  | 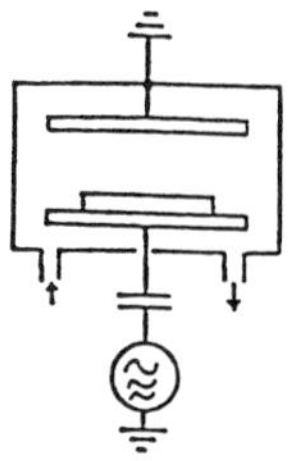 |
| Pressure [mbar] <br> Ion energy [eV] <br> Anisotropy | $10^{-1}-10$ <br> thermal <br> no | $10^{-1}-10$ <br> $<100$ <br> (yes) | $10^{-2}-10^{-1}$ <br> $10^2-10^3$ <br> yes |
| Advantages | No/low damage <br> High selectivity | Low damage <br> High etch rate and selectivity | Improved anisotropy <br> and linewidth control |
| Drawbacks | Passivation layer | Narrow process window <br> for anisotropic etch | Possible damage <br> Low etch rate |

| Method | Triode ion etch | Magnetron RIE | Reactive ion stream etching |
|---|---|---|---|
| Principle | 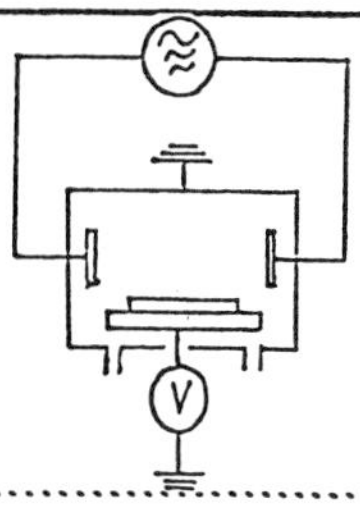 | 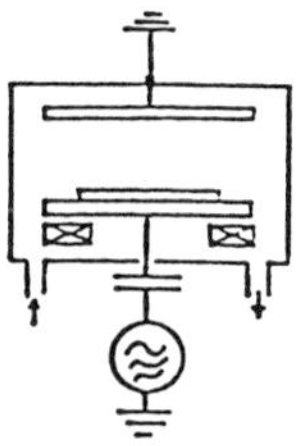 | 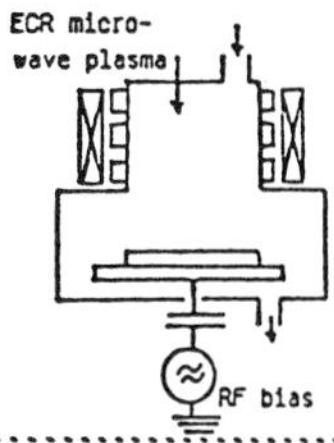 |
| Pressure [mbar] <br> Ion energy [eV] <br> Anisotropy | $10^{-1}-10$ <br> $10^2-10^3$ <br> yes | $10^{-3}-10^{-1}$ <br> $50-400$ <br> yes | $10^{-4}-10^{-3}$ <br> $<100$ <br> yes |
| Advantages | partial decoupling <br> of parameters | Good profile control, reduced <br> loading effect; high etch rate, <br> reduced damage | high selectivity <br> low damage |
| Drawbacks | possible damage | possible uniformity problems | New method |

Fig. 8. Dry etching techniques: range of process parameters and general advantages and drawbacks

Fig. 9.　Trench profiles after RIE in a $BCl_3/Cl_2$-chemistry.
　　　　left:　short mean free path for ions (sidewall roughness)
　　　　right: long mean free path for ions (pillar at the bottom of the trench)

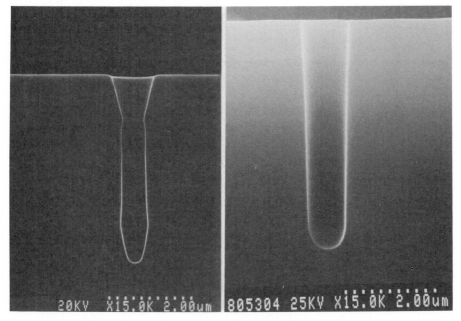

Fig. 10.　Trench profiles after triode ion etch.
　　　　left:　$CBrF3$-chemistry, inhibitor film not yet removed
　　　　right: $CBrF3/N2$-chemistry, inhibitor film removed;
　　　　　　　note the perfect hemispheric bottom of the trench

However two lithographic steps are necessary in the original approach in order to avoid $SiO_2$ - removal in the wide isolation region during etchback. In order to solve the narrow/wide isolation region problem with only one lithographic step a combined LOCOS/BOX approach has been proposed ((27), fig. 11). After groove formation and local oxidation a CVD oxide followed by resist coating smoothens the surface rather perfectly (step 4 in Fig. 11) since only short range planarization is necessary at this point of the process. A blanket anisotropic etchback with equal etchrates for resist and CVD-oxide leaves a rather plane surface where the nitride I layer serves as etch stop protecting the active regions. Uniformity and selectivity of etching is of course critical since the process has to continue until all nitride is cleared (fig. 12). Recently a modified planarization with combined RIE and chemical mechanical polish has been proposed (2).

Another application of planarization refers to the planarizing refill of small contact holes. Through contact holes electrical connection is made between upper conductor lines and underlying polysilicon or diffusion areas. Low contact resistance and high reliability under current stress are the major requirements. This led to the application of metal barrier layers underneath Al-based metallisation (e.g. TiN). In a conventional sputtering metallisation the edge coverage can be reduced to 10 - 20 % of plane thickness due to shadowing of the impinging atoms. An isotropic/anisotropic contact etch sequence (fig. 13 top) mitigates the shadowing to some extent. The posttreatment after contact etch which removes polymer residues and damaged substrate layers has been found to influence TiN barrier integrity drastically (fig. 13 bottom) due to sidewall roughness on a 100 nm-scale (28). The cracked barrier in fig. 13 (bottom right) leads to spiking of Al into single crystal silicon and a strong increase of junction leakage current, whereas the continous barrier after the anisotropic post-treatment prevents spiking.

It is clear from fig. 13 that this conventional contact hole metallisation is approaching its limits in the submicron regime. A complete metal filling of a contact hole can be achieved by conformal CVD-deposition of tungsten (29). In the example shown in fig. 14 a Ti/TiN-underlayer is used to achieve low contact resistance on $p^+$-and $n^+$-silicon and to improve adhesion of W. The tungsten etchback planarization was carried out in a $SF_6/O_2$-plasma (30) with a planarizing resist layer on top of W similar to the principle described above for isolation. Two major problems have to be solved in the etchback process: Firstly, avoidance of stringer residues along substrate steps which could lead to shorts between aluminum lines patterned later on in a Cl-based chemistry. Good planarization of the underlaying dielectric helps to overcome this problem. Secondly the socalled local loading effect. This would enhance the etchrate for tungsten plugs where the surrounding is already cleared compared to areas of delayed clearance. By proper adjustment of etchback process parameters the local loading effect can be avoided (30).

## ACKNOWLEDGEMENTS

For helpful discussions in preparing this manuscript we are indebted to. S. Schwarzl and H. Körner. Thanks go to S. Auer for simulation work performed for fig. 4 and to A. Krause for forwarding deep UV resist micrographs. The work of Mrs. Sapolska for typing the manuscript and preparing the figures is greatfully acknowledged. Part of the work described above has been supported by Federal Department of Research and Technology of the Federal Republic of Germany. The authors alone are responsible for the contents.

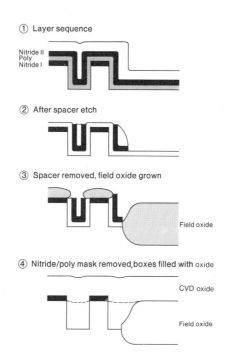

① Layer sequence

Nitride II
Poly
Nitride I

② After spacer etch

③ Spacer removed, field oxide grown

Field oxide

④ Nitride/poly mask removed, boxes filled with oxide

CVD oxide

Field oxide

Fig. 11. Process sequence for a combined BOX/LOCOS-isolation. Step 4 shows the planarized surface before and after etchback

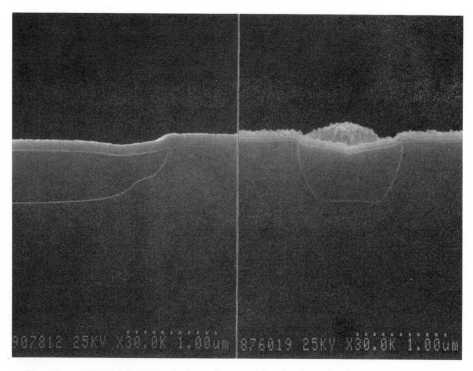

Fig. 12. LOCOS/BOX isolation after etchback planarisation.
Left: wide isolation region (made by LOCOS)
Right: narrow isolation region made by the trench isolation BOX

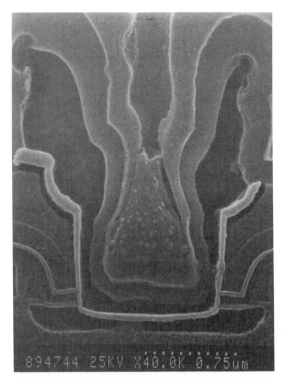

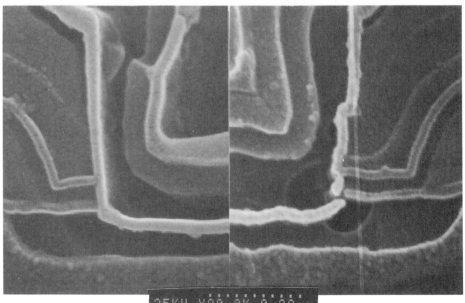

Fig. 13.  Top: contact hole after metallisation with Ti/TiN (thin bright layer)
and AlSiCu followed by passivation layers.
Bottom left: bottom edge of contact hole with continuous barrier
(anisotropic posttreatment)
Bottom right:  barrier with crack after isotropic posttreatment

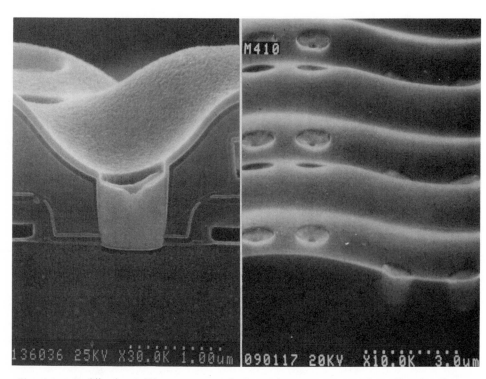

Fig. 14.   Refill of a 0.75 μm contact hole with tungsten. A TiN adhesion layer is used for W (left). Top view on an array of filled contact holes in reflowed BPSG dielectric (right).

# References

**(1)** A. Gutmann, J. Binder, G. Czech, J. Karl, L. Mader, D. Sarlette and W. Henke, SPIE 1990, Proceedings to be published

**(2)** B. Davari et. al., IEDM Tech. Dig. 1989, 61-64

**(3)** H. Fukuda, N. Hasegawa and S. Okazaki, J. Vac. Sci. Technol B7 (4), Jul./Aug. 1989, 667

**(4)** M. K. Templeton, C. R. Szmanda and A. Zampini, SPIE Vol. 771 (1987), 136

**(5)** P. Trefonas III, B.K. Daniels and R. L. Fischer, Jr., Solid State Technology, August 1987, 131

**(6)** S. Nakamura, K. Matsumoto, K. Ushida and Masaomi Kameyama, SPIE 1990, to be published

**(7)** A. Gutmann, A. Kleinhaus and W. Bade, Microelectronic Engineering 3 (1985), 329

**(8)** W. H.-L. Ma, SPIE Vol. 333, Submicron Lithography, (1982), 19

**(9)** "Photoreactive Polymers", A. Reiser, J. Wiley & Sons (1989), Chapter 7, "Deep UV Lithography"

**(10)** J. R. Sheats, Solid State Technology, June 1989, 79

**(11)** M. J. O'Brien and J. V. Crivello, SPIE Vol. 920 (1988), 42

**(12)** B. Reck et al., SPE Preprints 1988, 63

**(13)** J. W. Thackeray, G. W. Orsula, E. K. Pavelchek and D. Canistro, SPIE Vol. 1086, (1989), 34

**(14)** R. Schwalm, H. Binder, B. Dunbay and A. Krause, Polymers for Microelectronics, Conference, Tokyo, November 1989, Proceedings to be published

**(15)** J. Conway, Proc. KTI Microelectronics Seminar, Interface 88, 341

**(16)** D. Coyne, T. Brewer, Proceedings Kodak Interface '83, (1983), 40

**(17)** B. L. Draper, A. R. Mahoney and G. A. Bailey, J. Appl. Phys. 23 (1984), 1304

**(18)** T. Nogushi et al., SPIE Vol. 920 (1988), 168

**(19)** R. Sezi, M. Sebald and R. Leuschner, Polymer Engineering and Science, July 1989, Vol. 29, No. 13, 891

**(20)** J. M. Shaw, M. Hatzakis, J. Paraszczak and E. Babich, Microelectronic Engineering 3 (1985), 293

**(21)** J. P. W. Schellekens, Microelectronic Engineering 9 (1989), 561

**(22)** R. Sezi, R. Leuschner, M. Sebald, H. Ahne, S. Birkle and H. Borndörfer, Microelectronic Engineering (1990), to be published

**(23)**     private communication, S. Schwarzl

**(24)**     M. Engelhardt, S. Schwarzl
Proc. Symp. on Dry Process, Electrochem. Soc. Proc.
Vol. 88-7, 48 (1988)

**(25)**     M. Engelhardt, Proc. 15th Annual Tegal Plasma
Seminar, 53 (1989)

**(26)**     T. Shibata et al, IEDM Tech. Dig. 1983, 27

**(27)**     Ch. Zeller, F.X. Stelz, Proceed. 19th European Solid State Device
Research Conf., Berlin 1989, 135 - 138

**(28)**     private communication, H. Vogt

**(29)**     Pei-Ing Lee et al., J. Electrochem. Soc. Vol. 136, No. 7 (1989), 2108

**(30)**     J. Berthold, C. Wieczorek, Applied Surface
Science 38 (1989) 506

# QUANTUM WIRES AND DOTS : THE CHALLENGE TO FABRICATION TECHNOLOGY

S.P. Beaumont

Nanoelectronics Research Centre
Department of Electronics and Electrical Engineering
University of Glasgow, Glasgow G12 8QQ, UK

## INTRODUCTION

In these lectures I will argue that the fabrication of quantum dots and wires for optical applications presents the severest challenge to semiconductor nanotechnology. The dimensional requirements of these structures coincide with a variety of limitations in lithography and pattern transfer processes which are either fundamental or show few signs of being overcome. As a result, researchers in this field are seeking alternative, novel and in some cases bizarre ways of making quantum structures which raise important questions about our approach to device fabrication and bring microelectronics directly into the realm of molecular electronics and related issues.

Elsewhere in this volume, Weisbuch discusses the properties of quantum structures of 1- and 0-dimensionality and it is not necessary to repeat his findings here, suffice it to comment that the interest in this field is partly intellectual and partly directed towards the engineering of laser devices of narrow linewidth and lower threshold current, and classes of devices which utilise the predicted enhancements of oscillator strength and optical nonlinearities in quantum wires and dots.

## DIMENSIONAL REQUIREMENTS

It is important to realise that for optical purposes there are both upper and lower limits to the sizes of the structures we need to fabricate. This was pointed out clearly by Vahala [Vahala, 1988], and I use his arguments here. Vahala's paper referred solely to quantum dots, but the principles of his approach are applicable to quantum wires too.

By considering the confined states in a quantum sphere made from one semiconductor surrounded by a matrix of a second semiconductor, Vahala showed that:

a) There is a minimum size to the sphere below which there will be no confined electron states. The critical minimum diameter is given by the expression:

$$D_{min} = \sqrt{\frac{10\hbar^2}{m_e \Delta E_c}}$$

where $D_{min}$ is the minimum sphere diameter, $m_e$ the effective mass of the electron and $\Delta E_c$ the conduction band discontinuity between the sphere and the surrounding matrix. Note that the minimum diameter is dependent on the sphere material and also the matrix.

b) If the sphere is too small, the separation between the quantised states will be so small as to be undetectable in the presence of thermal broadening. If it is assumed that the separation between the states should be greater than 3kT where T is the ambient temperature and k Boltzmann's constant, then:

$$D_{max} \approx \sqrt{\frac{10\hbar^2}{6m_e kT}}$$

where $D_{max}$ is the maximum sphere diameter for detectable transistions: this critical dimension is dependent on the sphere material only.

Fig 1 shows the allowable dimensional range for quantum dots fabricated in the AlGaAs/GaAs and InGaAs/InP materials systems. The limits are shown for different ternary compositions and for a range of temperatures. The important message of this diagram is that, if we require quantum dot devices to operate at room temperature, then we have to fabricate structures whose dimensions are less than 20-30nm in size.

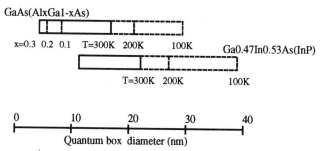

Figure 1. Allowable range of dot sizes for AlGaAs/GaAs
and InGaAs/InP quantum wells

## THE CHALLENGE TO FABRICATION TECHNIQUES

This dimensional requirement coincides almost exactly with a regime where even the best fabrication technologies begin to fail. This does not necessarily mean that structures of this size cannot be fabricated, but that there arises severe problems of fidelity, integrity uniformity and yield. In this section of the lectures I want to discuss the issues which limit the resolution of these technologies and consider how, and whether, they might tackled.

### The process cycle

First it is important to point out that we must consider not only the challenge to lithography, but also to pattern transfer techniques. Figure 2 illustrates the cycle of processes needed to fabricate a general 'device'.

III-V device fabrication normally starts with a growth process. One route to quantum dot fabrication usually starts with the growth of a quantum well layer by Molecular Beam Epitaxy. This is probably the highest quality process to be used in the whole fabrication cycle for quantum dots, although as we see later even it has deficiencies which give rise to artefacts. Subsequently the layer structure receives a variety of treatments: first, lithography is used to define a dot pattern. Next, that pattern is transferred into the substrate either by etching or by implantation. In some cases, a metal transfer layer may need to be deposited as a mask. All of these steps are imperfect in resolution to some degree. The following section treat these processes in turn.

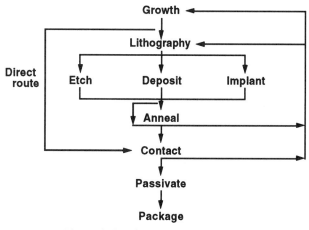

Figure 2. The 'Device' Process Cycle

## *Lithography*

First, we should consider which techniques for high resolution lithography are suited to fabricating our structures. Figure 3 shows the resolution capabilities of available methods. Clearly, optical lithography as a general purpose tool is eliminated: state-of-the-art equipment cannot achieve less than 0.3μm linewidth. However, the more specialised technique of holographic grating fabrication must not be ignored. In this process, interfering laser beams create a periodic intensity distribution which can be used to expose directly into photoresist. The periodicity of the grating depends on the wavelength of the laser light, the angle of convergence of the beams and the refractive index of the medium. Using this dedicated technique and introducing process bias, such as overdevelopment of the photoresist or overetching of the substrate it is possible to attain dimensions at which quantum effects can be observed in luminescence at very low temperature: we will return to these results later. Unfortunately the method does not offer the promise of reaching dimensions required for confinement at room temperature. It is worth bearing in mind, however, that the use of process bias has been used on many occasions to beat resolution limitations. Another example of the method is the use of metal shadowing followed by liftoff: in the very early days of nanostructure fabrication this approach was used to manufacture ultrasmall Josephson junctions using optical lithography.

If our goal is the attainment of sub-30nm dimensions, however, then we are left with just three possibilities. The obvious way to beat the diffraction limited resolution of optical

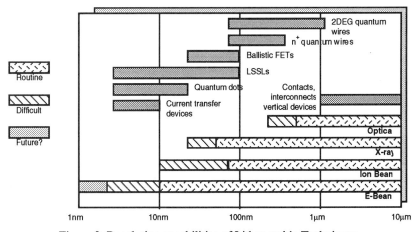

Figure 3. Resolution capabilities of Lithographic Techniques

lithography is to reduce the wavelength of the radiation employed. By recourse to x-rays, linewidths down to 17nm have been demonstrated. Unfortunately, x-ray lithography is currently only a contact printing process. There are no high-resolution x-ray lenses available at the moment, although some promising developments are in the offing. Since contact printing is a 1:1 process, the mask must be manufactured by a technique of writing arbitrary patterns with the same resolution as the final structure. X-ray lithography is thus a secondary process, potentially of great value for the mass production of quantum scale structures, but rather inflexible for most nanofabrication research.

This leaves us with electron beam and focussed ion beam lithographies. Assuming, for the moment, that we intend to write a pattern in resist by one of these techniques, we must first consider the resist process and the intrinsic limitations to resist resolution.

*Resists*

High resolution resists for electron, ion and x-ray lithographies are, in general, simple polymer backbones modified in some cases with aromatic sidegroups to improve dry etch sensitivity. They can be classified into two groups: positive and negative acting. In a positive resist, the exposed material is developed away. For a negative resist the reverse is true: the exposed material remains after development.

Positive resists work by a process of chain scission. Energy dissipated by absorption or inelastic scattering breaks up the polymer chain into fragments. As a result, the average molecular weight of the polymer decreases during exposure. The process of polymer fractionation in mixtures of solvent and non-solvent can then be used to selectively dissolve the shorter chain fragments. For example, poly(methyl methacrylate) or PMMA, the highest resolution positive resist, is developed typically in mixtures of methyl isobutyl ketone (MIBK), a solvent for this polymer, and isopropyl alcohol (IPA), a non-solvent. The critical weight below which the molecules will dissolve depends on the concentration of this mixture: for a 1:3 v/v solution the critical weight is about 10,000. The unexposed polymer typically has a molecular weight in the range $10^5$ to $10^6$.

Negative resists work in the opposite fashion. Short chain molecules in the unexposed resist are cross-linked on exposure into a network whose molecular weight is considerably larger than the precursor's. The fractionation process works in reverse too, the solvent/nonsolvent mixture dissolving away the unexposed, low molecular weight material.

The intrinsic resolution of a resist depends firstly on the nature of the interaction between the primary electrons (or ions) and the resist itself, that is on the delocalisation of energy from the initial trajectory of the primary beam, and secondly on the chemistry of the development process. Neither of these two factors are, in my view, properly understood but the first has received considerable, though limited, attention because of the well-known 'proximity effect', in which the exposure of adjacent pattern elements influence one another by the process of primary electron scattering. Electrons can be scattered in their passage through the resist and also backscattered into the resist again by the substrate. In experiments to determine resist resolution limits, both these effects can be reduced to negligible proportions by using thin resists and thin substrates, and by exposing the resist with high energy (50-100kV) electrons. Typically, 25-50nm of resist can still be uniformly spin-coated and processed, and can be easily supported by a similar thickness of silicon nitride or carbon. Even when these precautions are taken, and the resist is exposed by an electron beam of negligible diameter, the minimum linewidth produced in PMMA is approximately 10nm. This residual linewidth has been ascribed to the effects of secondary electrons generated by inelastic collisions of the primary beam, but the work of Rishton [Rishton, 1985] shows that this mechanism fails to account completely for the 10nm linewidth limit because the energy distribution of secondary electrons is strongly peaked at very low energy loss values where the range of secondary electrons in resist is small. Recent results from x-ray lithography also suggest that the distribution of secondary electrons is biased strongly towards small energy loss. Consequently we still have to devise a mechanism to explain the empirical resolution measurements. One possibility is the molecular size of the resist: although Broers has carried out experiments which show no relationship between resolution and molecular weight in PMMA, similar work by the Glasgow group on polystyrene, a negative resist, shows a definite worsening of resolution with increasing molecular weight. One must take care when drawing parallels between positive and negative resists because negative resists also swell during development as the developer penetrates the loose network of crosslinked molecules to form a gel: a process which, together with the molecular size effect, limits the resolution of negative resists to approximately 30nm. Positive resists are not believed to be prone to swelling: however, it might be that at the limit of resolution, the penetration of solvent into the sidewalls of the developing pattern causes swelling that obliterates the pattern. These solvent effects, and the possibility of energy delocalisation along the polymer chain, are potential mechanisms for the resolution limit.

*Particle beam optics*

Whilst the properties of the resist impose a fundamental limit on the resolution of the lithography process, the electron or focussed ion beam lithography equipment must be capable of forming a beam small enough to exploit that resolution. To understand the factors influencing the resolution of electron and ion beam machines we have to study a little bit of particle beam optics.

The conventional approach to calculating the resolution of a particle beam is to sum the contributions from:

a) the Gaussian demagnified image of the source

b) the spherical and chromatic aberrations of the lens which forms the probe

c) the effects of diffraction.

## Gaussian spot diameter

The contribution to the final spot diameter from the demagnified image of the source is obtained by invoking conservation of brightness

$$d_{gauss} = Dd_{source} = \sqrt{\frac{4I_p}{\pi\beta\alpha^2}}$$

where $I_p$ is the total current in the spot, $\beta$ the brightness of the source (current density per unit solid angle) and $\alpha$ is the angular half-aperture of the probe-forming lens.

## Spherical aberration

In a lens suffering from spherical aberration, rays making large angles with the optic axis are focussed more strongly than rays closer to the axis. The diameter of the disc of least confusion of the caustic of rays is given by

$$d_{sph} = \frac{1}{2}C_s\alpha^3$$

where $C_s$ is the coefficient of spherical aberration of the probe forming lens.

## Chromatic aberration

Chromatic aberration arises because the focal length of a lens depends on the primary particle energy. Thus the effect of any energy spread on the beam is to enlarge a point focus into a disc of diameter

$$d_{chr} = C_c\alpha\frac{\Delta E}{E}$$

where $C_c$ is the chromatic aberration constant of the probe forming lens and $\frac{\Delta E}{E}$ the fractional energy spread on the beam.

## Diffraction

Finally, diffraction at the aperture of the probe forming lens is a significant effect in particle lenses because apertures have to be small to limit the effects of spherical aberration. The diameter of the diffraction spot is given by:

$$d_{diff} = \frac{1.22\lambda}{\alpha}$$

where $\lambda$ is the relativistically-corrected de Broglie wavelength of the particles.

Total probe diameter

By convention, the total diameter of the beam is obtained by summing the above component diameters in quadrature:

$$d_p^2 = \frac{4I_p}{\pi\beta\alpha^2} + \frac{1}{4}C_s^2\alpha^6 + C_c^2\alpha^2\left(\frac{\Delta E}{E}\right)^2 + \frac{1.49\lambda^2}{\alpha^2}$$

Fig. 4 is a plot of $d_p$ against final aperture, $\alpha$, for a lens of having $C_s = C_c = 1cm$ and a source of brightness $10^5$ A/cm$^2$str at 30kV with a fractional energy spread of $10^{-5}$. A constant current of 10pA is assumed in the spot. These are typical values for an e-beam system working at small, though usable, beam current.

From the figure, it is clear that a minimum diameter is obtained at an aperture of approximately 10mrad. For the system in question the dominant components are the Gaussian and spherical aberration contributions. We can reduce the minimum diameter by reducing the beam current, thus reducing the Gaussian contribution, to the point where diffraction and spherical aberration dominate: this is typically the case in electron probe lithography machines. However, reducing beam current implies long exposure times, as resist is sensitive to the total amount of energy deposited in each pixel. If we want to preserve or even reduce exposure time, we have to substitute a source of greater brightness. Thus the highest resolution e-beam machines (usually converted scanning electron microscopes) use field emission guns with brightnesses greater than $10^8$ A/cm$^2$str, and are capable of forming diffraction/spherical aberration limited spots down to 0.5nm diameter. On the other hand, ion beam lithography machines are limited in performance by brightness and chromatic aberration, because the energy spread from an ion source is usually much greater than from an electron source, and the aberrations of ion lenses are significantly greater than those of electron lenses. Attempts to form ion probes on the order of 10nm diameter have so far failed, and there appears to be no prospect of success in the near future. To do so would require significantly brighter sources than those currently available, and either an improvement in lens aberrations (unlikely) or a smaller energy spread (requiring new source technology) to reduce chromatic aberration. In conclusion, the engineering of electron beams to exploit the resolution of available resists is straightforward, but to date the resolution of ion beam lithography is machine limited.

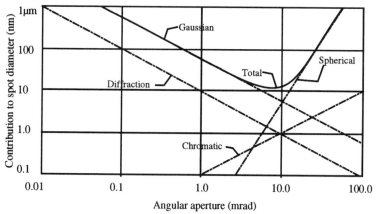

Fig 4. Contributions to total beam diameter from components
due to Gaussian image, diffraction and aberrations.

*Electron beam lithography*

Figure 5 shows a schematic section through an e-beam lithography machine. As discussed above, it is the properties of the final probe-forming lens which determine the size of the pattern writing spot, but in most systems other lenses have to be incorporated to contribute to the overall demagnification of the electron source into its Gaussian image in the final probe, and to control the probe current. The beam is scanned under computer control only over those elements in the pattern which require to be exposed: this technique is known as vector scanning, and within each element the beam writes in a zig-zag path to fill in the shape. Between each element the beam is blanked in order to prevent unwanted exposure.

Much electron beam nanolithography is carried out with converted scanning electron microscopes which have sufficient resolution to exploit the capabilities of PMMA resist, and sufficiently high energy to avoid series proximity effects. A new generation of commercial e-beam machines is emerging, however, which offer very high resolution and high energy. Such systems are engineered with high quality electron optics to allow large patterns to be scanned in reasonable time and with preserved resolution over the field, and laser-controlled interferometers which can stitch adjacent fields to an accuracy of a few nm.

*Focussed ion beam lithography*

Conventional ion implanation has not found a role in nanofabrication as such, though it may be exployed for device isolation. On the other hand focussed ion beam technology is beginning to be used for nanostructure fabrication for the purposes of:

a) direct (unmasked) implantation of dopants

b) direct (unmasked) etching by sputtering or stimulated chemical etching

c) direct deposition of metallic structures

d) lithography in resists

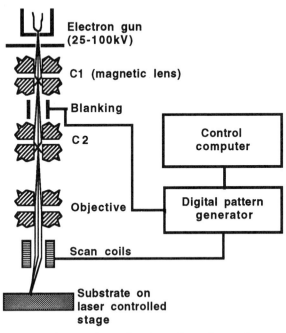

Figure 5. Schematic section through an electron beam lithography system.

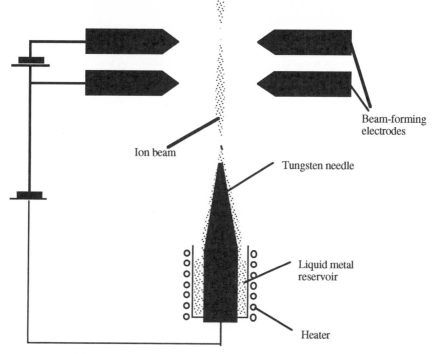

Fig 6. Schematic Diagram of a Liquid Metal Ion Source

Focussed ion beam equipment is conceptually similar to its electron beam counterpart but is of much more recent origin. Successful focussed ion beam systems awaited the development of a very bright source of ions which was realised in the liquid metal ion source (Fig 6). Here, a tungsten needle is wetted by capillary action from a molten reservoir of elemental (often Ga) or alloy (e.g. Au/Be/Si) material. The electric field applied to the tip extracts an intense beam of ions which can be focussed by a series of lenses. These are usually electrostatic because particles of large mass/charge ratio require magnetic fields too large to be generated by conventional ferromagnetic-cored magnetic lenses of a convenient size. Probes of 30nm minimum diameter can be formed in this way: the resolution is limited by the ion energy spread coupled with the chromatic aberration of the lens. Attempts have been made to improve on this resolution limit by trying to develop guns with reduced energy spread so far without success. The column of a focussed ion probe system usually incorporates a mass separator to select single from multiply charged ions or ions of different elements from alloy sources. The Au/Be/Si source, for example, allows one to switch between p- and n-type dopants for GaAs in the same writing task.

*Summary*

Conventional resist-based lithography is limited in resolution to dimensions of approximately 10nm by the properties of the resist itself. Electron beam lithography allows this resist limit to be reached, but focussed ion beam lithography is some way from this limit because of poorer technology: it has some useful features, however, in resistless lithography which might be advantageous in some schemes for nanostructure fabrication although it is difficult to conceive how sufficiently small and uniform structures could be fabricated to generate strong optical confinement.

### Pattern transfer

Faced with resist patterns limited to about 10nm feature size, we now address the issues of transferring this structure into semiconductor without loss of resolution.

*Liftoff*

The primary use for liftoff in quantum dot fabrication is to make a mask for etching or implantation. Liftoff is a very simple method of transferring (in principle) a positive resist structure into metal or any other material which can be deposited by thermal evaporation. The resist acts as a stencil for the evaporant.. By dissolving the resist after the coating step, unwanted material is washed away leaving the desired pattern on the substrate. This process is capable of defining 10nm metal wires but it can be difficult to pattern thick metal films because the resist gap tends to close laterally as the film is deposited. Moreover, the resolution of the pattern may be limited by the grain size of the metal: gold and aluminium, for example, have grain sizes much larger than the 10nm lithographic resolution limit. One must use fine grain alloys, such as gold-palladium, or refractory metals such as tungsten or molybdenum, if the quality of the lifted off structure is not to be dominated by grain size effects. Unfortunately, fine grain metals are usually either difficult to evaporate or to remove after processing.

*Wet and Dry etching*

Lifted off metal, or in some cases the resist pattern itself, can be used as a mask against etching. For nanostructures, wet etching is not normally employed. Wet etches are often difficult to control and they have a tendency to penetrate along the resist/substrate interface leading to ragged structures and rough surfaces.

For nanometre patterns dry etching is the preferred method. Here, etching is performed by ion bombardment and the material removed by physical sputtering, chemical reaction or combinations of both mechanisms. These are several 'flavours' of dry etching but all have in common the following:

a) the formation of a plasma in a noble and/or a reactive gas mixture

b) bombardment of the target sample by immersion in the plasma or by ions extracted from the plasma.

The important issues for the fabrication of quantum dots and other quantum-scale structures are:

(i) Resolution. It is very difficult <u>not</u> to achieve nanometre resolution unless the material is removed isotropically, as in plasma etching, and the resist is severely or completely undercut.

(ii) Low damage. Bombardment of a semiconductor by energetic heavy ions causes damage by either physical or chemical disruption. Figure 7 illustrates some of the processes which might cause such damage.

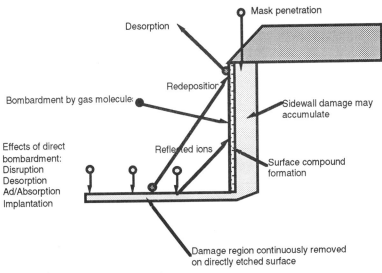

Fig 7. Illustrating some of the mechanisms of dry etch damage

Characterisation of damage and attempts to find low-damage etching processes are the focus of active research. Note that not only is the damage to the directly etched surface important: the sidewalls, whose surfaces often define the quantum confining structure, must not be detectably altered by etching. Sidewall damage is difficult to quantify, but characterisation has been attempted by measuring the cutoff widths of conducting wires [Thoms et al, 1989], and, as we shall see in the next section, by measuring the luminescence efficiency of quantum dots.

## LUMINESCENCE EFFICIENCY OF QUANTUM WIRES AND DOTS AND THE EFFECTS OF PROCESSING

Our own work in this topic has concentrated largely on the luminescence efficiency of quantum wires and dots fabricated by dry etching of GaAs/AlGaAs quantum well material. This is of technological importance, both because of the possibility that this method, followed by overgrowth, might be used for device fabrication, and because it is a severe test of fabrication technology and the difficulties that arise when making structures close to the limits of process resolution. It is also of basic scientific interest to understand the role of surfaces created by the etching step in addition to the search for quantisation effects.

The process adopted is illustrated in Fig 8. The starting material consisted of a variety of AlGaAs/GaAs quantum well samples, some grown by MBE and others by MOCVD. Typical MBE structures contained wells of increasing thickness with depth so that the luminescence from each well could be used as a depth probe. MOCVD samples, in contrast, contained a single well 10nm thick. Patterning was carried out first by using electron beam nanolithography at 50kV to define an etch mask. This was done either by lifting off metal films (eg NiCr) from a positive (PMMA) resist stencil or by using high resolution negative resist (HRN) directly as the etch mask. With the former technique, the mask was believed to be more uniform but it could not be certain that the metallisation would not interfere with the optical experiments. HRN, in contrast, was simpler to process, could be removed in an oxygen plasma prior to overgrowth and was transparent, but could not be guaranteed uniformly impervious to ion penetration especially close to its resolution limit of 40nm.

In our first experiments, MBE-grown layers were etched into quantum dots of various sizes down to 40nm diameter together with 100μm square mesas for process assessment and material uniformity control. Both metal and HRN masks were used and etching was carried out

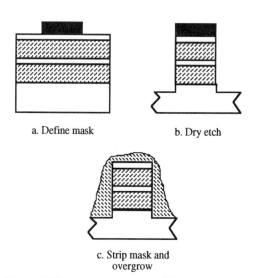

a. Define mask        b. Dry etch

c. Strip mask and
overgrow

Figure 8. Process steps used to fabricate quantum
dots by etching and overgrowth of quantum
well material

by silicon tetrachloride and methane/hydrogen reactive ion etching (RIE). Photoluminescence (PL) and photoluminescence excitation (PLE) experiments were carried out on many samples at 5K. No blue shifts or additional features in the PLE spectra due to quantum confinement were detected but, on average, the luminescence efficiency of all samples, scaled to the volume of emitting material, was independent of dot size and the processing technique adopted [Andrews et al, 1990]. Surface recombination appeared not to play a role in quenching the luminescence of the smallest of our structures at least within the accuracy of the measurement of luminescence intensity. This remarkable result agreed with earlier observations by Kash et al [1987], except that there was no convincing evidence of enhanced luminescence as was claimed in that work, but was seriously at odds with the results of Forchel [1988] who showed that in his quantum wire samples not only did excitonic diffusion to surface nonradiative recombination sites play a dominant role in quenching luminescence with shrinking dot size, but also that it was necessary to invoke the presence of a thick dead layer, caused by reactive ion etch damage, to fit his data to a mathematical model of the diffusion/recombination process. The dead layer issue is obviously important as it sets a limit to the smallest dot that can be fabricated without completely damaging the material, and also gives us information on the sidewall damage issue discussed above.

Since this early work, a number of authors have published data on luminescence efficiency, and a comparison is instructive. Although Clausen et al [1989], etching quantum dots with boron trichloride RIE, still have to invoke a dead layer to model their results, they show that its thickness depends on etch depth and is much thinner than the measurements of Forchel suggested. A more recent paper from Kohl, Heitmann et al, [1989] in which the first convincing evidence of quantum confinement in AlGaAs/GaAs quantum wires is presented, claims that optimal silicon tetrachloride reduces the luminesence efficiency of 70nm quantum wires by a factor of only 30, rather than orders of magnitude as reported by others. Fig 9 shows this published luminescence data together with out own data, normalised to the emission of unpatterned mesas and plotted against the smallest dimension of the structure.

Notice how the luminescence efficiency of structures fabricated in the same material system by the same basic process (electron beam lithography followed by dry etching) differ from one laboratory to another. Ones immediate reaction is to ascribe these differences to details associated with the processing. Clearly the thick dead layers needed to model Forchel's data are not present to the same degree in Arnot's wires, and even less so in Heitmann's. Presumably, if the dead layer is present at all, careful processing and optimisation of the etch process can minimise it's thickness. Exactly how the process can be tuned to achieve this result is still, to some degree, a black art: the measurements involved in the characterisation of quantum wires

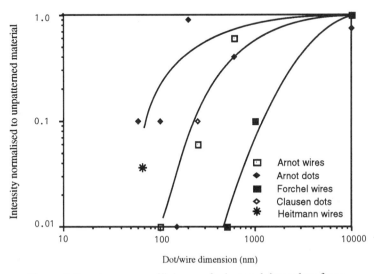

Figure 9. Luminescence efficiency of wires and dots taken from published data

and dots are difficult, so the mapping of dry etch parameter spaces to find optimum conditions involves a great deal of effort which appears not yet to have been expended.

In conclusion, our experiments have shown that the etching process need not create extensive optically dead sidewalls on quantum dots and that surface recombination can be inhibited by overgrowth. Etching and overgrowth remains a viable, though unattractive, candidate for device fabrication. Clearly there is a strong interplay between experiment and technology in this work and a full understanding of the results demands thorough characterisation of starting material (beyond simple PL measurements) as well as assessment of the processed structures themselves.

## CONCLUSIONS AND FUTURE TRENDS

Quantum dot fabrication poses a severe challenge to conventional semiconductor fabrication technology. Each step in a normal fabrication process appears to have its resolution limit within the size regime which we wish to explore when fabricating devices with optical confinement. Table 1 summarises the factors which limit resolution in conventional nanolithography:

Table 1. Resolution limits of conventional nanolithography

| Process | Resolution determined by: |
|---|---|
| Optical lithography | Diffraction |
| E-beam lithography | Resist / delocalisation of energy |
| Focussed ion beam lithography | Machine technology / resist |
| Liftoff | Metal grain-size, resist closure |
| Dry etching | Damage |
| Controlled/structured growth | Surface migration |

Further progress requires new approaches and these are already beginning to emerge. If one could overcome the resist resolution limits, the potential resolution of e-beam lithography could be exploited. Attempts are being made to do this using inorganic resists which are decomposed by the action of an intense electron beam [Kratschmer & Isaacson, 1986]. Features of the order or 2nm in size can be made visible in resist: the problem at the moment is to translate such small structures into useful materials. Direct growth on structured surfaces appears very promising [Kapon et al 1988, 1989]; using this method the first quantum wire lasers have been fabricated. Crystallite growth from glassy precursors or within zeolites and even the feeding of yeasts on cadmium salts are being attempted as more exotic routes, though it is difficult to envisage how such methods could integrate into a device process!

## REFERENCES

Andrews SR, Arnot H, Rees PK, Kerr TM and Beaumont SP, 1990, **J. Appl. Phys. 67**, 3472

Arnot HEG, Watt M, Sotomayor-Torres CM, Glew R, Cusco R, Bates J and Beaumont SP, 1989, **Superlattices and Microstructures 5**, 459-463

Arnot HEG, PhD Thesis, University of Glasgow, 1990

Clausen EM, Craighead HG, Worlock JM, Harbison JP, Schiavone LM, Florez L and van der Gaag B, 1989, **Appl. Phys. Lett. 55**, 1427

Forchel A, Meier HP, Maile BE and Germann R, 1988, **Festkörperprobleme 28**, 99

Kapon E, Harbison JP, Yun CP and Stoffel NG, 1988, **Appl. Phys. Lett. 52**, 607

Kapon E, Hwang DM and Bhat R, 1989, **Phys. Rev. Lett. 63**, 430

Kash K, Scherer A, Worlock JM, Craighead HG and Tamargo MC, 1987, **Appl. Phys. Lett. 49**, 1043

Kohl M, Heitmann D, Grambow P and Ploog K, 1989, **Phys. Rev. Lett 63**, 2124

Kratschmer E and Isaacson M, 1986, **J. Vac. Sci. Technol. B4**, 361

Rishton S, 1985] PhD Thesis, Glasgow University

Thoms S, Beaumont SP and Wilkinson CDW **J. Vac. Sci. Technol. B7**, 1823-1826, 1989

Vahala K, 1988, J. Quantum Elect. **QE22**, p1887

# CHEMICAL INTERFACES:
# STRUCTURE, PROPERTIES AND RELAXATION

A. Ourmazd

AT&T Bell Laboratories
Holmdel, NJ 07733, USA

## INTRODUCTION

Any finite system is delimited by interfaces. In this trivial sense interfaces are ubiquitous. However, modern epitaxial techniques seek to modify the properties of materials by the creation of interfaces. 'Band gap engineering', the attempt to tailor the electronic properties of semiconductors by interleaving many dissimilar layers is an extreme example of this approach. Technologically most advanced and thus most widely used are interfaces between lattice-matched, isostructural, crystalline systems, differing only in composition. These 'chemical' interfaces, are formed by the introduction of dopant impurities, or by the stacking of dissimilar materials. Chemical interfaces formed by the epitaxial growth of dissimilar materials are the subject of this article.

A fundamental tenet of modern epitaxy is the tailoring of properties through control of structure. We will thus outline how the 'structure' of a chemical interface may be defined and experimentally determined. The determination of structure is a major experimental challenge, inextricably bound with our understanding of the effect of structure on other properties. The article will thus contain a brief discussion of this vital link. We will conclude with an outline view of the way chemical interfaces, in reality systems far from equilibrium, can relax and how their relaxation mechanisms shed light on the fundamental properties of solids. Because the GaAs/AlGaAs system is the most technologically advanced, and for concreteness, we will illustrate the discussion by reference primarily to this system. Many of the concepts and experimental results, however, are of more general validity.

### DEFINITION OF A CHEMICAL INTERFACE

The (chemical) interface between two lattice-matched, isostructural materials can be uniquely defined on all length scales, provided each atom type occupies an ordered set of lattice sites. As an example, consider the GaAs/AlAs system. The interface is simply the plane across which the occupants of the Group III sublattice change from Ga to Al. The interfacial plane thus defined can in principle have a complex shape or 'waveform', with undulations ranging from atomic to macroscopic length scales. It is thus convenient to describe an interface in terms of its Fourier spectrum, by specifying the amplitude of the undulations as a function of their spatial frequency [1]. A 'white noise' interface, for example, would be characterized by a constant roughness amplitude over all possible length scales.

When a given experimental technique is used to investigate an interface, it provides information about the interfacial configuration within a certain frequency window, delimited on the high frequency side by the spatial resolution of the technique, and on the low frequency side by its field of view. For any single technique, this window spans only a small portion of the spatial frequencies needed for a realistic description of the interface. It is thus necessary to collate the data from a large variety of techniques to obtain a realistic picture of the interfacial configuration. This is a major challenge, because information from the atomic to the centimetre range, i.e. over eight orders of magnitude is required to provide such a description. Often, however, only specific characteristics, such as the optical or electronic properties of an interface are of concern, and knowledge of a limited range of frequencies is adequate. In the case of luminescence due to excitonic recombination in a quantum well, for example, roughness over the exciton diameter is of primary importance, while for charge transport, roughness at the Fermi wavelength is of concern.

The simple definition of an interface in terms of the location of the chemical constituents becomes inadequate when either of the parent materials is not chemically ordered, i.e., when some of the atom types are distributed randomly on a set of sites [1-3]. In the GaAs/$Al_xGa_{1-x}As$ system, for example, the second material is a random alloy. Thus, the Ga and Al atoms are distributed randomly on the Group III sublattice, subject to the constraint that the composition, averaged over a sufficiently large volume of the $Al_xGa_{1-x}As$ should correspond to the 'global' average x. This is schematically illustrated in Fig. 1, where random alloy $Al_{0.4}Ga_{0.6}As$ has been 'deposited' on an atomically flat GaAs surface, and the resulting structure viewed in cross-section. Each panel uses shades of grey to show the composition averaged over a given number of atoms perpendicular to the plane of the paper. Note that when only one atomic plane is shown, i.e.,no averaging is carried out, no continuous line can be drawn to contain only the Ga atoms. This illustrates that when one of the two constituent materials is a random alloy, an interface cannot be defined on an atom by atom basis. (Attempts to image interfaces by Scanning Tunnelling Microscopy be viewed in this light). Only as the 'thickness' over which the composition is averaged increases, does an isocomposition line approach the initial GaAs surface. These considerations apply generally, regardless of whether the interface is viewed in 'cross-section' as in Fig. 1, or in 'plan-view'.

## MICROSCOPIC STRUCTURE

The Transmission Electron Microscope (TEM) in its lattice imaging mode can in principle reveal the local atomic configuration of an interface. (The application of X-rays to the study of interfaces is not considered in this article). Conventional lattice imaging produces a map of the sample structure, and as such is not a useful probe of chemical interfaces. Here, we briefly describe the way the TEM may be used to obtain chemical information at near atomic resolution and sensitivity. We show that the combination of 'chemical lattice imaging' and vector pattern recognition allows one to quantify the composition of individual atomic columns.

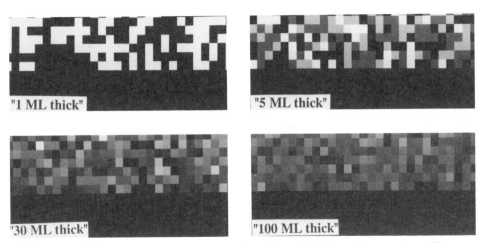

**Fig.1** *Schematic representation of the interface formed by depositing random allow Al$_{0.4}$Ga$_{0.6}$As on an atomically smooth GaAs surface (cross-sectional view). Only Group III atoms are shown. White represents pure Al, black pure Ga, other shades of grey intermediate compositions. In each case the composition of each atomic column is averaged over the 'thickness' of the sample. When the sample is only one monolayer thick, i.e. when there is no averging, no continuous line can be drawn to contain only the Ga (or only the Al) atoms, illustrating that an interface cannot bedefined on an atom-by-atom basis. Only as the thickness increases, does the interface become well-defined.*

## CHEMICAL LATTICE IMAGING

In the modern High Resolution TEM (HRTEM), a parallel beam of energetic electrons is transmitted through a thin sample to produce a diffraction pattern [4]. The phases and amplitudes of the diffracted beams contain all the available information. Part of this information is passed through an aperture, and is focused by the objective lens, causing the beams to interfere and produce a lattice image. In general, most of the reflections used to form a lattice image come about because of the lattice periodicity, and are relatively insensitive to the exact occupancy of the lattice sites. We name such reflections structural. However, certain reflections, such as the (200) in the zinc-blend structure, are due to chemical differences between the occupants of the different lattice sites, and contain significant chemical information [5-9]. Such chemical reflections are in general weaker than the 'strongly allowed' structural reflections, and the latter usually dominate the information content of lattice images. However, two factors, multiple scattering and lens aberrations, often considered disadvantageous of HRTEM, allow one to select and enhance the relative contribution of the weaker chemical reflections to lattice images [6,7].

When the electron beam enters a crystal along a high symmetry direction, a number of reflections are excited, exchanging energy among themselves as they propagate through the sample. To first order, this multiple scattering process may be viewed as the scattering of electrons from the undiffracted beam to each reflection, and their subsequent return. Structural reflections are strongly coupled to the undiffracted beam, and thus exchange energy with it rapidly as they propagate through the sample.

This energy exchange is slower for the more weakly coupled chemical reflections. Because of this 'penellosung' effect, at certain sample thickness a chemical reflection can actually have a larger amplitude than its structural counterpart. Appropriate choice of sample thickness can thus enhance the chemical information content of the lattice image. Moreover, the severe aberrations of electromagnetic lenses mean that the objective lens is essentially a bandpass filter, whose characteristics can be controlled by the defocus [4,6]. Thus, judicious choice of defocus allows the lens to select, and thus further enhance the contribution of the chemical reflections to the image.

To obtain chemical lattice images of compound semiconductor heterointerfaces in practice, advantage is taken of the chemical sensitivity of the (200) reflections. The sample is viewed in the d orientation, and the (200) (chemical) and (220) (structural) reflections are used to form an image. The sample thickness and lens defocus are chosen to maximize the change in the frequency content of the lattice image across the interface. Thus the chemical information in the sample is encoded into periodicity information in the lattice image with maximum sensitivity [7]. It is important to note that chemical lattice imaging techniques can be applied not only to semiconductors, but to any material in which the compositional change occurs by the substitution of atomic species on an ordered set of lattice sites. High $T_c$ superconductors [8], and metallic systems such as $Ni/NiAl_3$ [10] are examples of other materials amenable to chemical lattice imaging.

## ATOMIC CONFIGURATION OF CHEMICAL INTERFACES

Figure 2a is a structural lattice image of an InP/InGaAs interface of the type widely used to investigate the nature of chemical interfaces. It is of course true that even these structural images reveal, to some extent, the chemical change across the interface through the change in the background intensity. The question is whether this sensitivity is sufficient for such images to reveal the atomic details of the interfacial configuration. Fig.2b is the same as Fig.2a, except that the line marking the position of the interface is removed. The interface position and configuration are now less clear. This indicates the limited chemical sensitivity of structural images. Fig. 2c is a chemical lattice image of the same atom columns, obtained under optimum conditions for chemical sensitivity [7]. The InP is represented by the strong (200) periodicity (2.9 Å spacing), and in the InGaAs by the (220) periodicity (2 Å spacing). Clearly, the interface is not atomically smooth, the roughness being manifested as the interpenetration of the (200) and (220) fringes.

While this demonstration establishes the inadequacy of structural lattice images to reveal the interfacial configuration, it does not necessarily imply that all semiconductor heterointerfaces are rough. However, when GaAs/AlGaAs interfaces, shown to be of the highest quality by a variety of techniques [11-13] are examined by chemical imaging, atomic scale roughness is still observed [14]. This is revealed in Fig.3 as an interpenetration of (200) and (220) fringes at the interface, albeit a more subtle level. However, the evaluation of lattice images by visual inspection, though common, is subjective and unsatisfactory. We describe below a digital pattern recognition approach, which quantifies the local information content of lattice images, allowing their quantitative evaluation [14,15].

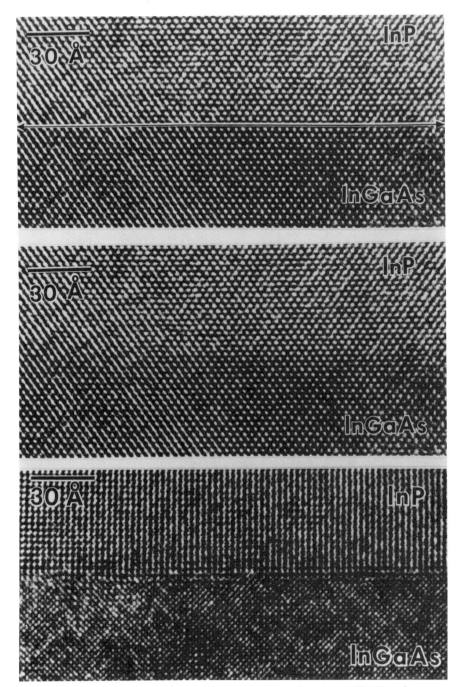

**Fig.2** *a) <110> (structural) image of an InP/InGaAs interface. The line draws attention to the interface.    b)   Same image without line.   c)   Same area of interface imaged along <100> under chemically sensitive conditions.    Note the interpenetration of InP (200) and  InGaAs  (220) fringes, indicating  interfacial roughness.*

**AlGaAs**      **GaAs**      **5.6Å↕**

**growth direction** ➡

**Fig.3**   *Chemical lattice image of GaAs/Al$_{0.37}$Ga$_{0.63}$As quantum well produced after two minutes of growth interruption at each interface. Careful inspection reveals interfacial roughness.*

## QUANTIFICATION OF LOCAL INFORMATION CONTENT OF IMAGES

It has already been established that the information content of a unit cell of a chemical image is directly related to the composition of a region of closely similar cross-section in the sample [15]. Thus, the local analysis of a chemical lattice image is equivalent to a highly local compositional analysis of the sample. Local analysis of an image can be most conveniently affected by real-space rather than Fourier analysis [15]. Real-space analysis proceeds with the examination of the information content of a unit cell of the image, which is most efficiently carried out by pattern recognition techniques. Here we describe a simple vector pattern recognition approach that efficiently quantifies the information content of a lattice image.

The task is carried out in several steps. First, perfect models, or templates are adopted from simulation or developed from the data, which serve to identify the ideal image of each unit cell type. When the template is extracted from experimental image, it is obtained by averaging over many unit cells to eliminate the effective noise (Fig.4). Second, an image unit cell of a particular size is adopted, and divided into an n x n array of pixels, at each of which the intensity is measured. Typically n ~ 40, and thus 1600 intensity measurements are made within each unit cell. Third, each unit cell is represented by a multidimensional vector, whose components are the $n^2$ intensity values measured in the cell. The ideal image unit for cell for each material is now represented by a template, which in turn is represented by a vector $R^t$. For example, the ideal image unit cells of GaAs and Al$_{0.37}$Ga$_{0.63}$As are characterized by the two vectors $R^t_{GaAs}$ and $R^t_{Al0.37Ga0.63As}$, respectively (Fig.4).

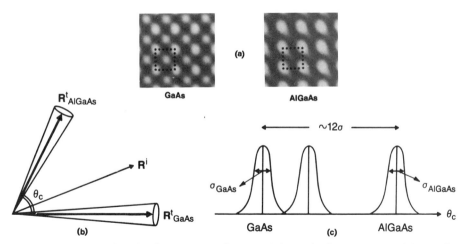

**Fig.4** *a)   Averaged, noise-free images of GaAs (left) and $Al_{0.37}Ga_{0.63}As$ (right). The unit cells used as templates for pattern recognition are the dotted 2.8 Å squares.*
*b)   Schematic representations of the template vectors $R^t_{GaAs}$ and $R^t_{Al0.37Ga0.63As}$, the distribution of $R_{GaAs}$ and $R_{AlGaAs}$ about them, and an interfacial vector $R^i$.*

Next, the amount of noise present in the experimental image is deduced from the angular distributions of the real (that is, noisy) unit cell vectors $R_{GaAs}$ and $R_{Al0.37Ga0.63As}$ about their respective templates. The noise in Fig.3 is such that, away from the interface, the $R_{GaAs}$ and $R_{Al0.37Ga0.63As}$ form similar normal distributions centred around their respective template vectors $R^t_{GaAs}$ and $R^t_{Al0.37Ga0.63As}$. The standard division $\sigma$ of each distribution quantifies the noise present in the images of GaAs and $Al_{0.37}Ga_{0.63}As$ (Fig.4). A unit cell is different from a given template, with an error probability of less than 3 parts in $10^3$, if its vector is separated from the template vector by more than 3 $\sigma$. The centres of the distributions for the GaAs and $R_{Al0.37Ga0.63As}$ unit cells shown in Fig.3 are separated by 12 $\sigma$, which means that each unit cell of GaAs and $R_{Al0.37Ga0.63As}$ can now be correctly identified with total confidence. A representation of the results of the vector pattern recognition analysis of Fig.3 is shown in Fig.5. This image is divided into $2.8 \times 2.8 \text{Å}^2$ cells, each of which is placed at a height representing the angular position of its vector. The light shaded cells in the foreground and the dark cells at the rear of the plate lie within 3 $\sigma$ of $R^t_{GaAs}$ and $R^t_{Al0.37Ga0.63As}$ respectively, while the other shades represent 3 to 5, 5 to 7 and 7 to 9 $\sigma$ bands [14,15]. (An alternative representation places a surface over the image, with the composition being represented as the height of the surface. This latter representation can be used to show the composition and hence confinement potential profile of a quantum well).

Fig.5 is essentially a quantitative chemical map of the $GaAs/Al_{0.37}Ga_{0.63}As$ interface of Fig.3. The height of a unit cell represents its composition, and the shade variation represents statistically significant changes in composition over and above random alloy statistics. This representation quantitatively displays the composition at each Group III atomic column ~ 15 atoms thick. The compositional change from 0 to 0.37 corresponds to changing a column of 15 Ga atoms to one containing 10 Ga and 5 Al atoms. This produces a 12 $\sigma$ signal. Thus, the replacement of a single Ga atom

**Fig.5** *Three-dimensional representation of the analysed lattice image of $Al_{0.37}Ga_{0.63}As$ grown on GaAs after a 2 min. interruption. The unit cells are 2.8 Å squares. The height of each cell represents the angular position of its vector R with respect to the template vectors, which are about 12 σ apart. The shades of grey used at the front and rear of the picture mark those cells which fall within 3 σ of GaAs and $Al_{0.37}Ga_{0.63}As$ templates respectively. Other shades of grey represent 3-5, 5-7 and 7-9 σ bands. In these regions, the Al content of each unit cell is intermediate between GaAs and $Al_{0.37}Ga_{0.63}As$, with confidence levels given by normal statistics.*

with Al gives rise to a 2.4 σ signal, and can be detected with 90% confidence. This argument is somewhat naive, in that it ignores the environment of the replaced atom. Nevertheless, it demonstrates that Fig.5 is essentially a map of the composition at near-atomic resolution and sensitivity. Although luminescence shows the interface of Fig.5 to be of the highest quality [11-13], it is clear that its atomic configuration is far from 'ideal'. The transition from GaAs to $Al_{0.37}Ga_{0.63}As$ takes place over 2 mono-layers (MLs), and the interface contains significant atomic roughness.

## MESOSCOPIC AND MACROSCOPIC STRUCTURE

Chemical maps of the kind shown in Fig.5 reveal the microscopic configuration of chemical interfaces. However, an adequate description of an interface requires information over a much wider range of length scales. Due to their limited field of view, direct microscopic techniques cannot be used to establish interfacial configuration over mesoscopic (micron) or macroscopic (millimetre) length scales. To make further progress, it is necessary to use indirect methods to probe the interfacial configuration. Such techniques attempt to determine the interface structure through its influence on the properties of the system, such as its optical or electronic characteristics. Fundamental to this approach is the premise that it is known how the structure effects the particular property being investigated. In practice this is rarely the case. 'Indirect' experiments thus face the challenge of simultaneously determining the way a given property is affected by the structure and learning about the structure itself.

Because a direct correlation is thought to exist between the structure of a quantum well and its optical properties, luminescence techniques have been extensively applied to investigate the structure of semiconductor interfaces [11-13,16-18]. In luminescence experiments the carriers optically excited across the band gap form excitons and recombine, often radiatively. The characteristics of a photon emitted due to the decay of a single free exciton reflect the structural properties of the quantum well averaged over the region sampled by the exciton. In practice, the observed signal stems from a large number of recombining exitons, some of which are bound to defects. The challenge is to extract information about the interfacial configuration form luminescence data, which represent complex weighted averages of the well width and interfacial roughness sampled by a large collection of excitons.

The photoluminescence (PL) spectrum of a typical single quantum well ~50Å wide grown under standard conditions, consists of a single line ~ 4.5 meV wide, at an energy position that reflects the well width and the barrier composition [11]. This linewidth is significantly larger than that of a free exciton in high quality 'bulk' GaAs (~0.2 meV), indicating additional scattering, presumably partly due to interfacial roughness. When the growth of the layer is interrupted at each interface, and the next layer deposited after a period of tens of seconds the PL spectrum breaks into two or three sharper lines each ~1.5 meV wide. This reduction in the PL linewidth is ascribed to a smoothing of the interfaces during the growth interruption. The two or three lines obtained from a single quantum well are often assigned to excitonic recombination in different regions of the quantum well under the laser spot, within each of which the well is claimed to be an exact number of atomic layers thick, ie. it is ascribed to recombination within 'islands' over which the interfaces are atomically smooth.

On the basis of this model, a quantum well of nominal thickness n in fact consists of regions (islands), within each of which the thickness is exactly (n-1), n or (n+1) monolayers (MLs), between which the interfacial position changes abruptly by 1 ML. These islands have been claimed to be as large as 10 $\mu$m in diameter [18], but are generally thought to lie in the micron range [12,13], and in any case to be much larger than the exciton diameter (~ 15 nm). A consequence of this model is that the luminescence peak separations must necessarily correspond to the difference in the energies of excitons that recombine in regions of the well differing in thickness by exactly 1 ML. In practice, the splittings rarely correspond exactly to ML changes in well width [19]. Departures from 'ML' values are generally ascribed to experimental uncertainties in determining the peak positions to fluctuations in the composition of the material, to impurities, or to exotic configurations of atomically smooth interfaces.

Quantitative chemical imaging, however, shows these interfaces to be atomically rough. The apparent contradiction is resolved by recalling that each experimental technique probes the interfacial configuration over a limited range of spatial frequencies. Thus, luminescence and chemical imaging results simply reveal different parts of the interfacial roughness spectrum. But the concept that luminescence shows interfaces to be smooth at the atomic scale has become so entrenched that it is important to examine the luminescence data carefully to determine whether they can indeed sustain the 'atomically smooth model'.

Warwick et al [1] have investigated quantum wells characterized by luminescence to have atomically smooth interfaces [11-13,18]. In their experiments, the laser spot was moved over samples held at 2 K, and a series of PL and PL Excitation (PLE) spectra obtained from sets of neighbouring points on each sample containing single quantum wells. At each point, PL and PLE data were also recorded form the AlGaAs barrier material and the GaAs buffer layer. In this way, the energies of the photons emitted from the well, the local mean Al content of the barrier, and the effects of residual stress were directly measured at many points on each sample.

Fig.6 shows the separation between the '21 ML' and the '22 ML' PL peaks obtained from a quantum well nominally 60Å thick, versus the position of the laser spot [1]. Note that the observed splitting varies in magnitude by nearly 40%. As indicated by the error bars, this variation is an order of magnitude larger than effects expected from the measured compositional fluctuations in the barrier and variations in the residual stress. Similar features are observed in PLE. The variations in the PL and PLE peak separations are clearly incompatible with the model of an atomically smooth and abrupt interface. This illustrates the inadequacy of attempting to describe the complex waveform of an interface in terms of a single island size, i.e. a -function in the frequency spectrum. However, the compilation of information over the necess-

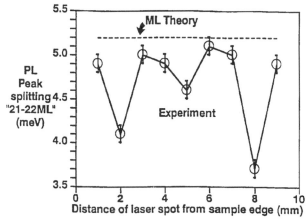

**Fig.6**  *Photoluminescence line splitting from a single nominally 60 Å thick quantum well vs position on the wafer. The splitting expected from atomically smooth interfaces and a thickness change from 21 to 22 ML is shown by the dashed line. Note the large (40%) variations in the measured splittings [1].*

ary frequency range, extending from the nm to the cm range is a formidable task. Warwick et al have thus attempted a qualitative compilation of the data from a variety of techniques as follows [1].

From the roughness revealed by chemical lattice imaging [14], they surmise the presence of 'significant' roughness at the atomic scale. On the other hand, the relatively sharp luminescence lines indicate 'little' roughness at wavelengths comparable with the exciton diameter. The occurrence of several sharp PL lines when a $\sim 100\,\mu$m diameter laser spot is used to excite luminescence, the observation of 'islands' in cathodoluminescence [12,13,18], and the observed variations in the PL peak splittings indicate 'substantial' low frequency roughness. The resultant roughness spectrum and interfacial configuration are schematically shown in Fig.7.

If such qualitative remarks can be taken seriously, the interfacial roughness spectrum is at least bimodal, with a minimum in the vicinity of the exciton diameter. This may be no accident; growth procedures have been so optimized as to give sharp luminescence lines, thus pushing roughness away from length scales comparable with the exciton diameter. While this may indeed be appropriate for materials intended for optical applications, it may not be suitable for transport experiments and electronic devices, where roughness at the Fermi wavelength must be minimized [20].

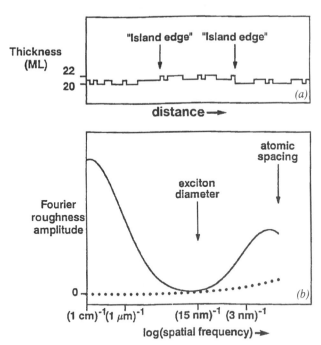

**Fig.7** *Schematic representation of the well thickness (a) and its Fourier transform (b). Roughness due to random alloy fluctuations is shown dotted in (b) but is too fine to be represented in (a) [1].*

## RELAXATION OF CHEMICAL INTERFACES

Because semiconductor multilayers are becoming increasingly familiar, it is easy to forget that they are highly inhomogeneous systems far from equilibrium. On crossing a modern GaAs/AlGaAs interface, the Al concentration changes by several orders of magnitude in a few lattice spacings. It is thus scientifically interesting and technologically important to investigate stability of chemical interfaces against interdiffusion. In most semiconductors, the modest diffusivities of point defects obviate substantial relaxation at room temperature. However, an interface can relax during thermal annealing, in-diffusion of dopants, or ion-implantation, processes necessary for devices fabrication. Rather than reviewing the extensive literature available, [21], we outline the new insights that emerge when the chemical relaxation of individual interfaces is studied at the atomic level.

Using the quantitative chemical mapping techniques described above, it is straightforward to make sensitive measurements of interdiffusion at single interfaces. The composition profile across a given interface is measured in two pieces of the same sample, one of which has been annealed (Fig.8). Starting with the initial profile and using the diffusion coefficient D as free parameter, the diffusion equation is solved to fit the final (annealed) profile, thus deducing D as a function of temperature and interface depth. In this way diffusivities as small as $10^{-20}$ cm$^2$/s can be measured in regions only $10^{-19}$cm$^3$ in volume [22,23]. Fig.9 is an Arrhenius plot of D vs 1/kT for C-doped GaAs/Al$_{0.4}$Ga$_{0.6}$As interfaces grown at three different depths beneath the surface, and subsequently annealed in bulk form. Remarkably, the magnitude of the interdiffusion coefficient, and the activation energy for intermixing change strongly with depth. Since this behaviour is observed both in the GaAs/AlGaAs and the

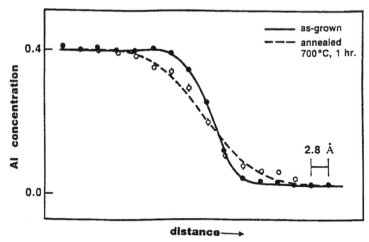

**Fig.8**  *Composition profiles of a C:GaAs/AlGaAs interface at a depth of 300Å, as-grown (solid line), and after 700C, one hour anneal (dotted line). One standard deviation error bars are shown. Each measurement refers to a single atomic plane, and is obtained by averaging the composition over 30 image unit cells on each plane [23].*

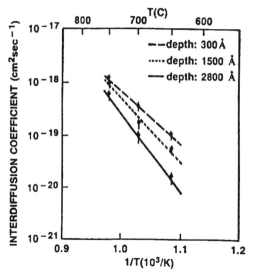

**Fig.9** *Arrhenius plot of the interdiffusion coefficient D at C:GaAs/AlGaAs interfaces grown at three different depths and subsequently annealed [23]*

HgCdTe/CdTe systems [22-24], it must be concluded that the depth-dependence of the interdiffusion coefficient is a general effect. This has been shown to be related to the injection of point defects from the sample surface during the anneal. In particular, interdiffusion in these systems is assisted by the presence of native point defects (interstitials or vacancies), whose concentrations are often negligible in as-grown samples. For interdiffusion to occur, such native defects must arrive from the surface during the anneal. The interdiffusion coefficient is a sensitive function of the concentration of these defects at the particular interface studied, and thus can be used to investigate the microscopies of native point defect diffusion in multilayered systems. Indeed, it should be possible to measure separately, the formation energy and migration energy of a given native defect (interstitial or vacancy) as a function of its charge state.

Returning to interdiffusion, two important points emerge. First, the interdiffusion coefficient varies strongly with depth. Thus a measurement of this parameter is meaningful only if it refers to a single interface at a known depth. Second, it follows that the interface stability is also depth-dependent. Thus the layer depth must be regarded as an important design parameter in the fabrication of modern devices. This effect assumes additional importance when interdiffusion is also concentration dependent, leading to strong intermixing at very low temperatures [25].

Since interdiffusion in high quality multilayers requires the assistance of a native point defect, the measurement of the relaxation of a chemical interface is tantamount to recording the arrival and passage of point defects through the interface. It should thus be possible to record the passage of energetic particles, such as ions implanted into a solid, by measuring the intermixing at a series of chemical interfaces in their path. The intermixing can be characterised by measuring the chemical profiles of individual atomic planes crossing the interface [26]. Thus each atomic plane crossing a

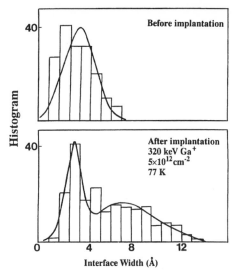

**Fig.10** *Histogram showing distance over which the composition changes from AlAs to GaAs along individual planes crossing an (AlAs on GaAs) interface. (a) Before implantation. (b) After implantation with 5 x 10¹² cm⁻² 320 keV Ga⁺ ions at 77 K. This dose corresponds to an average of one ion per 2000 Å² area of the interface. The appearance of a second peak is attributed to the passage of single ions through isolated areas of the interface [26].*

GaAs/AlAs interface is characterised by the distance (or width) L over which the composition changes form GaAs to AlAs. Fig.10 shows a histogram of the widths L for a collection of 300 planes across an (AlAs on GaAs) interface before and after implantation by 320 keV Ga+ ions at a dose of 5 x $10^{12}$ cm$^{-2}$ at 77 K [26]. The combination of the extremely low dose (one ion per 200 Å$^2$), and the observation of a bimodal distribution of widths after implantation indicate that certain areas of the interface have not been pierced by any ions, while other regions have been pierced by only one passing ion. The histogram of Fig.10 essentially shows that it is possible to detect the passage of single energetic ions through an interface in the same way that layers of photographic emulsion are used to record the passage of cosmic radiation.

The study of individual chemical interfaces at high spatial resolution and chemical sensitivity thus brings an unexpected bonus; chemical interfaces can be used to record the passage of point defects, be they due to the low energy elementary excitations of a solid (Frenkel defects), or high energy implanted ions. This offers an exciting window into the world of microscopic processes with both scientific and technological implications.

**SUMMARY**

With the exception of the $Si/SiO_2$ system, interfaces of the highest perfection, and thus widest application are those between lattice-matched pseudomorphic, crystalline solids, differing only in composition. Two concepts are fundamentally important in the understanding of such chemical interfaces. First, the definition of an interface is most conveniently affected in terms of its roughness spectrum, where the amplitudes of the interfacial undulations are specified as a function of their spatial frequency. Second, when one of two materials forming the interface is random alloy, the interfacial configuration cannot be defined at the atomic level.

Experimentally, it is essential to realise that any technique probes only a small part of the roughness spectrum. This 'window' is delimited on the high frequency side by the spatial resolution of the technique, and on the low frequency side by the field of view. Moreover, a technique may possess an intrinsic length scale: the exciton diameter for luminescence, the Fermi wavelength for transport, which plays a crucial role in determining the wavelength of the interfacial roughness that is most sensitively probed. To gain a realistic impression of the interfacial configuration, information over a wide range of frequencies is needed. It is thus necessary to collate the data produced by a variety of techniques. The description of an interface in terms of an 'island size' is an attempt to replace the real roughness spectrum essentially by a single frequency component. This is too naive to be realistic.

Chemical interfaces and multilayers are systems far from equilibrium, able to relax through interaction with point defects. This allows them to be modified by suitable processing for device applications. Equally importantly, chemical interfaces can be used to record the passage of point defects, providing a microscopic view of the processes that govern the elementary structural excitations of solids.

M. Bode, Y. Kim, J.A. Rentschler, D.W. Taylor, C.A. Warwick, J. Cunningham, M.W. Hong and R. Malik have been intimate collaborators in this work; I am grateful to them for permission to reproduce and discuss their results. I have also benefitted from valuable discussions with A. Bourret, L.C. Feldman, W. Schröter and J. Shah.


## ACKNOWLEDGMENT

This review is reproduced with the permission of the Materials Research Society and contains work originally published in MRS Bulletin Vol XV(9) p58 (1990) .

## REFERENCES

[1]    C.A. Warwick, W.Y. Jan, A. Ourmazd, T.D. Harris and J. Christen, Appl. Phys. Lett., in press.
[2]    M. Thomsen and A. Madhukar, J. Cryst. Growth, 84, 98, (1987).
[3]    S.B. Ogale, A. Madhukar, F. Voillot, M. Thomsen, W.C. Tang, T.C. Lee, Y.Kim and P. Chen, Phys. Rev. B36, 1662, (1987).
[4]    See e.g. J.C.H.. Spence, Experimental High Resolution Electron Microscopy, Oxford Univ. Press, New York, 1980).
[5]    P.M. Petroff, A.Y. Cho, F.K. Reinhart, A.C. Gossard, and W. Wiegmann, Phys. Rev. Lett. 48, 170 (1982).
[6]    A. Ourmazd, J.A. Rentschler, and D.W. Taylor, Phys. Rev. Lett. 57, 3037, (1986).

[7]   A. Ourmazd, W.T. Tsang, J.A. Rentschler, and D.W. Taylor, Appl. Phys. Lett., 50, 1417, (1987).

[8]   A. Ourmazd, and J.C.H. Spence, Nature 329, (1987).

[9]   A. Ourmazd, J. Cryst. Growth 98, 72, (1989).

[10]  J.M. Penisson and A. Bourret, private communication.

[11]  C.W. Tu, R.C. Miller, B.A. Wilson, P.M. Petroff, T.D. Harris, R.F. Kopf, S.K.Sputz, and M.G. Lamont, J. Cryst. Growth 81, 159 (North-Holland 1987).

[12]  R.C. Miller, C.W. Tu, S.K. Sputz, and R.F. Kopf, Appl. Phys. Lett. B 49, 1245, (1986).

[13]  P.M. Petroff, J. Cibert, A.C. Gossard, G.J. Dolan, and C.W. Tu, J. Vac. Sci. & Technol., B5, 1204 (1987).

[14]  A. Ourmazd, D.W. Taylor, J. Cunningham and C.W. Tu, Phys. Rev. Lett. 62, 933, (1989).

[15]  A. Ourmazd, D.W. Taylor, M. Bode and Y. Kim, Science 246, 1572, (1989).

[16]  C. Weisbuch, R. Dingle, A.C. Gossard, and W. Wiegmann, Solid State Comm. 38, 709 (1981).

[17]  M. Tanaka, H. Sakaki, and J. Yoshino, Jap. J. Appl. Phys. 25, L155, (1986).

[18]  D. Bimberg, J.. Christen, T. Fukunaga, H. Nakashima, D.E. Mars, and J.N.Miller, J. Vac. Sci. Technol., B5, 1191, (1987).

[19]  J.C. Reynolds, K.K. Bajaj, C.W. Litton, P.W. Yu, J. Singh, W.T. Masselink, R. Fischer and H. Morkoc, Appl. Phys. Lett. 46, 51 (1985).

[20]  H. Sakai, T. Noda, K. Hirakawa, M. Tanaka, and T. Matsusue, Appl. Phys. Lett. 51, 1934 (1987).

[21]  See e.g., D.G. Deppe and N. Holonyak, Jr., J. Appl. Phys. 64, R93, (1988).

[22]  Y. Kim, A. Ourmazd, M. Bode, and R.D. Feldman, Phys. Rev. Lett. 63, 636, (1989).

[23]  Y. Kim, A. Ourmazd, R.J. Malik, and J.A. Rentschler, Proc. Mat. Res. Soc., in press.

[24]  L.J. Guido, N. Holonyak, Jr., K.C. Hsieh, and J.E. Baker, Appl. Phys. Lett. 54, 262(1989).

[25]  Y. Kim, A. Ourmazd, and R.D. Feldman, J. Vac. Sci., Technol. A8, 1116, (1990).

[26]  M. Bode, A. Ourmazd, J.A. Rentschler, M. Hong, L.C. Feldman, and J.P. Mannaerts, Proc. Mat. Res. Soc., in press.

# CAPACITANCE-VOLTAGE PROFILING OF MULTILAYER
# SEMICONDUCTOR STRUCTURES

J.S. Rimmer, B. Hamilton and A.R. Peaker

The Centre for Electronic Materials
University of Manchester Institute of Science & Technology
P.O. Box 88, Manchester M60 1QD, U.K.

## INTRODUCTION

C-V profiling has traditionally found its importance in the determination of semiconductor doping profiles and, more recently, to provide estimates of heterojunction band offsets [1]. It has been recognised [2] that a profile of the free electron concentration $n(x)$ is obtained from C-V techniques rather than that of the doping profile $N_d(x)$. These are not necessarily the same even for the case of very shallow donors. Indeed, they will be different whenever the measured electron concentration is not uniform. It has also been shown [3] that C-V profiling viewed as a measurement of the free carrier concentration is itself inexact if $n(x)$ varies appreciably over a distance less than a Debye length. The measured apparent carrier concentration $\bar{n}(x)$ then differs from both $n(x)$ and $N_d(x)$. There is also still some uncertainty in the use of C-V profiling to determine heterojunction band offsets due to the effects of a finite donor depth [4] and of interfacial charge distributions [5].

Computer simulations have been found useful in overcoming some of these difficulties. One calculates theoretically what form of C-V relationship would result from some assumed dopant distribution or heterojunction band structure, and then compares the simulated results with the experimental curves. One can then guess a better assumed profile based on the remaining discrepancies. The time taken to arrive at a satisfactory conclusion is ultimately governed by the number of iterations necessary to match the simulated and experimental curves. This usually depends on the complexity of the structure. In order for C-V profiling to remain a rapid assessment technique, it is therefore necessary to approach the problem of modelling these structures efficiently so as to reduce computer run times to a minimum, and to include the effects of relevant physical processes commonly found in real semiconductor materials.

The way in which a modelling approach is applied to interpreting the C-V profiles of more complex structures is probably best illustrated by describing a specific example. In this paper we describe work on double heterostructure confined systems.

We also demonstrate the importance of using other experimental probes in parallel with C-V methods in order to reinforce our understanding of these multilayer structures. Many of these complementary techniques are described in this volume.

## C-V SIMULATION APPLIED TO DOUBLE HETEROSTRUCTURES

The existence of charged mid-gap states or recombination centres at the interfaces of semiconductor heterostructures is a problem of long standing. Estimates of the level of charge at a single interface have recently become possible by examining the apparent carrier concentration profile obtained from C-V measurements [1]. The GaAs/AlGaAs interface has been investigated extensively using C-V and DLTS (Deep Level Transient Spectroscopy) techniques including spatial profiling of the identified deep levels. However, there seems to be little agreement in the measured characteristics of the traps between different reports [6,7,8]. Measurements of interface recombination velocities in double heterostructures (DH) have been used as an indication of interface quality [9,10,11] and also to determine the effect of thin prelayers on the active region interfaces [12]. These DH structures must necessarily have thin optically active regions so that interface recombination will be the dominant process [13], if prelayers are included then the resulting structures may be relatively complicated. It is, therefore, clear that the levels of interface charge in these multilayer structures cannot be obtained by direct examination of the C-V profile as in the case of a single interface so that determining the sheet charge density presents a considerable problem.

Computer simulation of C-V profiles has also become established as an investigative technique. However, until recently, computer models have been restricted to a single interface and have often been beyond the capabilities of desk-top computers or at the very least, required considerable run times in the case of multilayer structures. Their use has been limited to investigating the inherent limitations of conventional C-V profiling [1,2,3,4] or as a check against band offset and interface charge measurement at single interfaces [14,15]. However, a new, fast C-V simulator package of software has now been developed [16] which runs on a desk-top computer and can provide simulated profiles of multilayer structures in a matter of minutes. This simulator can be used to determine the level of charge at each interface in a series of GaAs/AlGaAs DH structures, allowing us to see quantitatively the variation of charge with each successive interface. Interface recombination velocities can also be determined from time resolved photoluminescence measurements and a correlation found with the magnitude of charge at the active region interfaces. In addition, the parallel application of these techniques has revealed important clues as to the nature and characteristics of the recombination centre and their possible role in low dimensional devices.

## SAMPLE STRUCTURE

The structure of the samples used to illustrate these methods is shown in figure 1. All were produced in a VG 90 MBE system on $n^+$ GaAs substrates after the growth of a $0.5\mu$m GaAs buffer which was silicon doped to $10^{18}$ cm$^{-3}$. Substrate temperature was maintained at 680°C and growth was under arsenic-stable conditions as determined by RHEED. The DH regions were silicon doped to $2 \times 10^{16}$ cm$^{-3}$ and

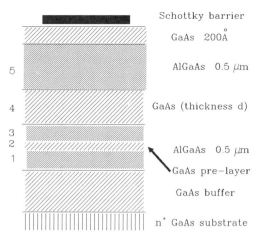

*Figure 1.    Heterojunction sample structure.  The numbers on the left correspond to the regions marked in Figure 2.  The active region thickness, d is given in Table 1.  The gallium arsenide prelayer when present is 200 Å thick.*

the layer thicknesses were designed so that the depletion region formed by a Schottky barrier could be swept through all the hetero-interfaces.  The aluminium mole fraction was determined by PL to be 36%.  The layer thickness for one sample was determined by TEM and the others were inferred from the relative growth times.  Two of the samples were grown without prelayers and these are identified along with the active region thicknesses for all samples in table 1.

*Table 1.  Active layer thicknesses and pre-layers in DH samples.*

| Sample No. | d (Å) | Pre-layer |
|:---:|:---:|:---:|
| 37 | 2708 | no |
| 43 | 2708 | yes |
| 44 | 1190 | yes |
| 47 | 650 | yes |
| 48 | 1950 | yes |
| 49 | 1300 | yes |
| 64 | 1300 | no |

## EXPERIMENTAL METHODS

### Interface Charge

Using gold Schottky diodes it is possible to measure C-V profiles by depleting through the interfaces for samples 37, 44, 47, 49 and 64.  Computer simulated profiles were fitted to the latter four samples.  An example of experimental and simulated curves (for sample 49) is shown in figure 2.  The curves were fitted by using the shallow dopant concentration determined from the flat band regions of the experimental curves and assuming a conduction band offset of 0.27 eV, which corre-

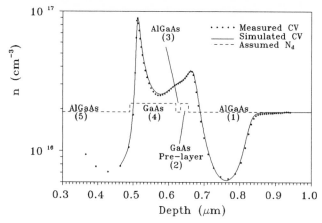

*Figure 2.    Measured and simulated C-V profiles for sample 49.*

sponds to 60% of the total band offset. A fit was obtained by adjusting the levels of interface charge only, the other parameters were kept constant. The results for the four samples are shown in figure 3. It is important to note that the best fit was obtained by considering negative interface charge, independent of bias, and in fact, no feature was observed corresponding to the emission of electrons as the Fermi level crosses the trap level [17] in all but sample 64. This sample yielded the highest level of charge at the inverted interface and gave a weak feature in the profile which would correspond to an activation energy of 1 eV, as shown in figure 4.

For sample 37, the active region was wide enough to allow the CV profile to fall to its flat band condition between the interfaces. It was, therefore, possible to apply Kroemer's method [1] to determine the charge at the inverted interface.

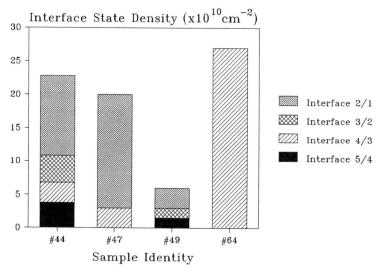

*Figure 3. Distribution of charged state sheet density between the interfaces of DH structures as determined by CV simulation.*

## Recombination Velocity

The recombination velocity $S$ is related to the effective minority carrier lifetime $\tau_{eff}$ as determined from photoluminescence decay measurements by the equation [13]

$$\frac{1}{\tau_{eff}} = \frac{1}{\tau_b} + \frac{S_1 + S_2}{d} \qquad (1)$$

where $\tau_b$ is the bulk recombination lifetime and $d$ is the layer thickness with $d < < L$, the minority carrier diffusion length. Since the recombination velocity for samples with prelayers could be widely diferent from those without [12], the average recombination velocity and bulk lifetime were therefore obtained by determining $1/\tau_{eff}$ as a function of $1/d$ for the samples with prelayers only. It was then assumed that the quality of the bulk GaAs in the active region was more consistent from sample to sample than the interface quality so that $\tau_b$ could be considered approximately constant over the whole series, including those samples without prelayers. The value obtained for $\tau_b$ (183 ns) was then used in equation 1 to determine $S_1 + S_2$ for each individual sample. The resulting values for the four samples used in the C-V measurements are plotted against the charged state density for the active region interfaces in figure 5.

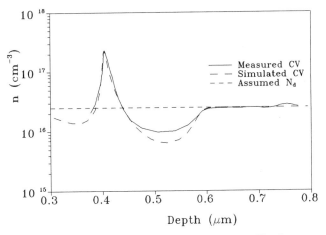

Figure 4.    Measured and simulated C-V samples for #64. The feature corresponding to emission from deep level at 1 eV can be seen on the far right of this plot..

The charge density at the inverted interface of sample 37 was determined using Kroemer's method [1] and the recombination velocity was determined in the usual way. The results fit the relationship of figure 5 almost exactly and in fact would place a point virtually on top of that for sample 64 although it has been omitted for clarity.

## DISCUSSION

The use of this C-V simulator software allows the level of charge to be quantitatively assessed for each interface in multilayer structures. It is difficult to estimate the experimental error in this determination but by keeping the adjustable parameters consistent between samples it is most probable that this will be at worst systematic and will, therefore, have minimal effect on the final analysis of the data. In addition, the technique will only measure the *net* interface charge which may include contributions of opposite polarity from, for example, accumulations of shallow donor impurities. There were also some discrepancies between the experimental and simulated C-V curves in the HJ depletion regions which we believe to be due to dynamic effects associated with the test signal frequency [18] as the simulator is based on a d.c. model. However, the observed negative charge states tie in with the recombination velocities in an intuitively expected manner. Furthermore, sample 37 gave a result which was consistent with data from the other samples even though the interface charge was determined by an independent and generally accepted method. This too tends to suggest that any errors in the levels of charge deduced by simulation will be within acceptable limits.

In figure 3, it is clear that the majority of the charge resides at the earliest grown interfaces. Sample 64 has no prelayer and in this case the entire charge is determined to be at the first grown (inverted) interface. It can be seen that the total charge summed over all interfaces is similar for each sample except 49 which had a much reduced level. The PL data for this sample indicated a much higher quality AlGaAs in relation to the other samples and this suggests that the interface quality is related to the bulk AlGaAs quality. This supports the hypothesis that the interfaces may act as a gettering plane for impurities. It is also seen in figure 5 that the interfacial recombination velocity bears a direct relationship to the number of charged states at the active region boundaries from which it is deduced that these states are the dominant recombination centres.

The recombination will be rate limited by minority carrier capture provided $n\sigma_n > p\sigma_p$, where $\sigma$ is the capture cross section, and this should be true for low excitation. If it is assumed that this is so and that the species at each interface is similar then the recombination velocity $S$ can be expressed as:

$$S = \sigma_p \, v_{th} \, N_{st} \qquad (2)$$

where $v_{th}$ is the thermal velocity and $N_{st}$ is the sheet state density. Applying this relationship to the data of figure 5 yields a value close to $10^{-15}$ cm$^2$ for the minority carrier capture cross section.

We therefore have some evidence from the C-V and PL data that there is an accumulation of a deep acceptor at the interfaces of these heterostructures with an energy about 1eV below the conduction band edge. The magnitude of the estimated hole capture cross section is consistent with the hole 'seeing' a negatively charged centre and it is this negatively charged state which is observed in the C-V measurements. These indications of the characteristics of the defect have arisen as a direct result of the parallel application of C-V and optical techniques.

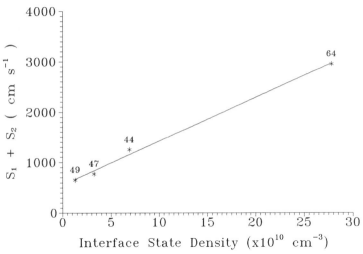

*Figure 5. Recombination velocity plotted against interface state sheet density.*

Further work is presently underway to investigate the defect using DLTS and photocurrent techniques. Preliminary results using MCTS (Minority Carrier Transient Spectroscopy) have been encouraging but the data is difficult to analyse due to the compositional non-uniformity of the heterostructures and the sheet nature of the defect. It is intended to study a new series of samples with hole injection from p-type regions.

## CONCLUSIONS

C-V simulation may be realistically used to determine interface charge in multi-layer heterostructures. This approach has led to a quantitative appraisal of the gettering of impurities at the earliest grown interfaces in the GaAs/AlGaAs system. The parallel application of optical techniques has clearly shown a direct relationship between the magnitude of these states and interfacial recombination and has indicated a dependence of the interface quality on the bulk AlGaAs quality.

In addition, it has been estimated that the state has an energy approximately 1eV below the conduction band edge and a hole capture cross section of approximately $10^{-15}$ cm$^2$. It therefore seems likely that the defect may be further characterised by minority carrier capture or transient methods.

## REFERENCES

[1]    Kroemer H.,Chien Wu-Yi, Harris Jr. J.S., Edwall D.D., (1980), Appl. Phys. Lett. 36, 295.

[2]    Kennedy D.P., Murley P.C., and Kleinfelder W. (1968), IBM J. Res. Dev. 12, 299.

[3]     Johnson W.C. and Panousis P.T. (1971),IEEE Trans. Electron.. Dev
        *ED-18,65.*

[4]     'tHooft G.W., and Colak S., (1986)Appl. Phys. Lett. 48 (22), 1525.

[5]     Okumura H., Misawa S. and Yoshida S., (1986), Surf. Sci. 174, 324.

[6]     Matsumoto T., Bhattacharya P.K., Ludowise M.J., (1983),
        Appl. Phys. Lett. 42 (1), 52.

[7]     Mc Afee S.R., Lang D.V., Tsang W.T., (1982), Appl. Phys Lett. 40 (6), 520.

[8]     Ohno H., Akatsu Y., Hashiyume T., Hasegawa H., Sano N.,
        Kato H., Nakayama M., (1985), J. Vac. Sci. Technol. B3 (4), 943.

[9]     Nelson R.J., and Sobers R.G., (1978a), Appl.Phys.Lett.32 (11),761.

[10]    Sermage B., Pereira M.F., Alexandre F., Beerens J. Azouay R.,
        Kobayashi N., (1987), Gallium Arsenide and Related Compounds, Heraklion.

[11]    Sermage B., Alexandre F., Lievin J.L., Azoulay R., El Kaim M.,   Le Person H., and
        MarzinJ.Y., (1985), GalliumArsenideandRelated Compounds,Hilger,345.

[12]    Dawson P., and Woodbridge K., (1984), Appl. Phys. Lett. 45 (11),1227.

[13]    Many A., Goldstein Y., Grover N.R., (1971), Semiconductor Surfaces,
        North-Holland, Amsterdam, 259.

[14]    Leu L.Y. and Forrest S.R., (1988), J. Appl. Phys., 64 (10), 5030.

[15]    Watanabe M.O., Yoshida J., Mashita M., Nakanisi T., Hojo A., (1985), J. Appl.
        Phys., 57 (12), 5340.

[16]    Rimmer J.S., Hawkins I.D., Hamilton B., Peaker A.R., (1989) ,"III-V
        Heterostructures for Electronic/Photonic Devices", MRS Proceedings Vol. 145, 75.

[17]    Kimerling L.C., (1974), J. Appl. Phys. 45 (4), 1839.

[18]    Subramanian S., Vengurlekar A.S., (1988), J. Appl. Phys. 64 (3), 1552.

# CONCEPTS AND APPLICATIONS OF BAND STRUCTURE ENGINEERING
# IN OPTOELECTRONICS

M. Jaros

Physics Department, The University
Newcastle upon Tyne NE1 7RU, UK

BEYOND THE PARTICLE IN A BOX PICTURE

The concept of electronic structure engineering is as old as
semiconductor physics. Soon after the discovery of binary compound
semiconductors such as GaP and GaAs it was realised that suitable
solid mixtures of these materials could be used to alter the
magnitude of the forbidden gap. The emergence of the optic cable
provided a particularly strong impetus for such experimentation and
as a result a new technology came into being of quaternary alloys
involving elements of Ga, In, P, As etc. grown on high quality InP
substrates (see, for example, Pearsall, 1982). Concurrently with
the development of new alloys it has become possible to grow high
quality epitaxial layers of technologically important semiconductors
of almost arbitrary (and well controlled) thickness. In an alloy
the band gap is changed by alterations of the alloy composition.
There is a simple linear relationship between the alloy composition
and band gap. A similar linear relation exists between the gap and
the lattice constant. Semiconductor heterostructures offer a
different way of achieving changes in band gap. When a thin layer
of, say, GaAs is sandwiched between layers of $Ga_{1-x}Al_xAs$, the larger
gap material (the alloy) acts as a simple potential barrier for
electrons at the bottom of the conduction band of GaAs (e.g. Capasso
and Margaritondo, 1988). The electron levels in GaAs are shifted as
a result of the confining potential and their position with respect
to the bulk band edges of GaAs can be estimated from the particle in
a box model (Jaros, 1989).

It is apparent that in the heterostructure the band gap
engineering has been taken a step further. However, in our example,
we have tacitly assumed that we can account for the relevant change
in the electronic structure merely by invoking the bulk properties
(Fig. 1) of the constituent materials, just as we have done in order
to describe the electronic structure of semiconductor alloys. In
fact, it turns out that in this case, at least for $x < 0.45$ when the
alloy semiconductor remains a direct gap material, our assumptions
are well justified. Let us examine these assumptions in some
detail. We know that at the band edges of a perfect semiconductor,
say GaAs, the electron wave function resembles a standing wave whose
period is given by the separation between atoms. Recall, that in a
one dimensional example this wave function is cos $(\pi/a)$ where a is
the lattice constant. What happens to such a wave function at the
interface with $Ga_{1-x}Al_xAs$? Since the lattice constant of these
materials is so similar we can take it to be for our purposes the

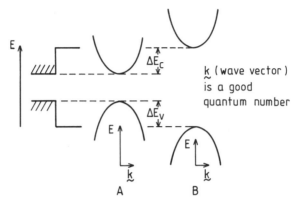

Fig.1. A band offset for a GaAs quantum well.
A) ... GaAs, B) ... $Ga_{1-x}Al_xAs$.   $\Delta E_c$ and $\Delta E_v$ are the
conduction and valence band offsets respectively.

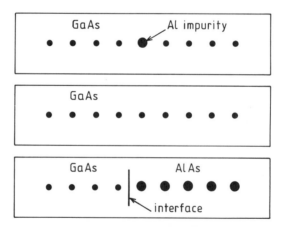

Fig.2. A schematic illustration of the impurity model (top), and
of the analogous way a superlattice can be constructed out
of a bulk GaAs crystal by replacing atoms of Ga with those
of Al at the relevant sites.

same.  The difference between the crystals of GaAs and $Ga_{1-x}Al_xAs$
"seen" by the GaAs wave function is then represented by the
difference between the potential of Al atom and that of Ga atom
which we have to replace (Fig. 2) in order to "create" the $Ga_{1-x}Al_xAs$
out of GaAs (i.e. for x = 0.5 we replace on average every second Ga
with Al).  What happens at the interface to the GaAs wave function
therefore depends on the strength of the difference between the
atomic potentials of Al and Ga.

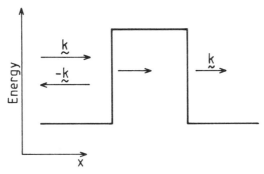

Fig. 3.  A sketch of reflections at a rectangular barrier.  k is the
wave vector of the incoming wave.

The way we look upon the problem in Fig. 2 reminds us of the
similarity between the familiar impurity problem (Jaros, 1982) in
bulk materials and that of the quantum well and superlattice.  In
the impurity problem we also start with host (bulk) perfect crystal
wave function and ask what is the difference between the impurity
atom and the host crystal potentials in the vicinity of the impurity
site.  If this difference or "impurity" potential seen by the bulk
crystal waves can be approximated by a screened Coulomb potential,
we can find a hydrogenic series of additional localised levels in
the gap.  If this impurity potential is stronger because of the
larger difference between the impurity and host atoms, particularly
in the region of atomic core, then the localised levels lie deeper
in the gap.  On the other hand, if the impurity potential is very
weak it may not be capable of producing bound states.  In such
circumstances we can only expect the crystalline waves to interact
weakly with the impurity and at best be reflected at the site
without a significant scattering taking place.  If we  proceed to
replace not one but many atoms of Ga with those of Al, we may
generate a layer of say AlAs out of GaAs.  We can guess the degree
of interaction at the interface by looking at the impurity problem
which is usually well known from the studies of extrinsic bulk
properties.  For example, it is known that an isolated Al impurity
in GaAs does not give rise to localised levels in the gap of GaAs.

We know that the strength of the atomic potential is also reflected in the magnitude of the band gaps and in the form of the bulk band structure of a semiconductor in general. The band structures of GaAs and $Ga_{1-x}Al_xAs$ for small x (e.g. 0.3) are very similar except for the small difference in the band gap at k = 0. There is therefore every reason to think that the difference between the atomic potentials of Al and Ga is also insignificant i.e. that the shape of the solution of the Schrödinger equation in the range of the atomic potential of Al is nearly the same as that in the range of Ga. Hence the waves from GaAs merely change their direction (reflect) at the interface but retain the same wavelength. This is precisely what one assumes when considering a reflection and matching of plane wave functions at a rectangular barrier. (Fig. 3.) Similarly, when a layer of GaAs is confined from two sides by $Ga_{1-x}Al_xAs$, the waves inside the GaAs well are reflected from the barriers and form standing waves which associate with the confined discrete levels in the well.

This examination should draw our attention to two important aspects of the problem that are normally left out of the discussion. Firstly, we can express quantitatively the statement that the two constituent semiconductors forming a quantum well heterostructure must be "similar" for the particle in a box model to hold good; the similarity means that the potential difference between the atoms facing each other across the interface must not disturb the characteristic wavelength (the length of the wave vector k) of the original bulk wave function. Secondly, it is borne in mind that the wave function of a confined state in the GaAs well must retain the nodal character given by the atomic separation in bulk GaAs (Fig. 4) since it is constructed from an incoming and reflected bulk crystal wave function of GaAs which can be identified in a simple one dimensional model as cosine and sine standing waves. This behaviour is superimposed upon the familiar cosine or sine like form which results from the near total reflection at the confining barrier. Hence the confinement supplies only an envelope or modulating function and the total wave function in the quantum well should really be written as a product of the envelope function whose argument depends on the well width $l$ and the rapidly varying bulk component whose argument is $\pi/a$, i.e. $\cos (\pi/l) \cos (\pi/a)$ where $a \ll l$. Of course, the position of the confined levels with respect to the bulk band edge depends only on the confinement effect and the rapidly varying bulk component of the wave function is normally ignored. We have decided to recover this full form of the wave function here because it will help us to understand the approximations on which the particle in a box model is based, and also point the way of going beyond the limits of this model.

BAND STRUCTURE EFFECTS

The Effect of Atomic Potential

We are now in a position to ask what would happen if the difference between Al and Ga were not so small. Take, for instance, the case of CdTe and HgTe quantum well heterostructure. Again, the lattice constants of the two bulk materials are so similar that we can ignore the difference. However, the atom of Hg is very different from that of Cd. As a result the solution of the Schrödinger equation in the region of the atomic potential of Hg must be expected to take a very different shape from that in the range of the atomic potential of Cd. Let us consider a wave function $\psi$(CdTe) of CdTe and move towards the interface with HgTe. In order to

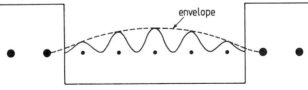

Fig.4. Top: a sketch of the electron charge density representing
a quantum state at the bandedge of a bulk crystal of GaAs.
Bottom: the form of the confined charge density in a GaAs
quantum well, showing the modulating influence of the
confinement effect.

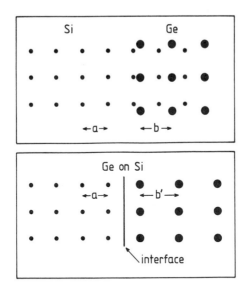

Fig.5. Construction of a Si-Ge superlattice by replacing Si atoms
with those of Ge in the Ge layers; top: with bulk
coordinates (e.g. b), bottom: Ge coordinates assuming the
lattice constant a of the silicon substrate in the
interface plane and expanding the atomic separation (b')
in the direction perpendicular to the interface

change $\psi$(CdTe) into $\psi$(HgTe) (match them at the interface as required by quantum mechanics) we need more than just the waves reflected and transmitted at the same wavelength. We must assume that the difference between these bulk wave functions $\psi$(HgTe) and $\psi$(CdTe) is really a wave packet containing wave vectors of different lengths. In the language of the impurity model developed in the above paragraphs we could expect an isolated Hg impurity in an otherwise perfect bulk crystal of CdTe to behave as a "deep" impurity, i.e. to give rise to highly localised states or wave packets containing many different momentum (k) components. It has been shown that this is indeed the case.

We can rephrase this intuitive picture of localisation in HgTe-CdTe microstructures in a mathematical form. Suppose we decide to express the wave function of HgTe at the bottom of the conduction band (at k = 0) in terms of the solutions of the Schrödinger equation for bulk CdTe. In fact there is a theorem in quantum theory which says that such solutions always form a complete set so that any well defined function can be expressed as a linear combination of $\psi$(CdTe) taken at all possible points in the wave vector space. Hence we can formally write $\psi$(HgTe)$_{k=0}$ = $\Sigma_k$ $A_k$ $\psi$(CdTe)$_k$. This is the form of a wave function that is actually obtained in full scale calculations. Of course, if the difference in Hg and Cd potentials was insignificant then only one coefficient in this expansion namely $A_{k=0}$ would be needed. Clearly, the larger the difference between Hg and Cd potentials the stronger the mixing of waves of different wave vectors into the wave function across the interface. In the case of a GaAs quantum well considered earlier we only expect a reflection to occur hence the expansion will contain only enough coefficients to recover the cosine envelope function.

The effect of mixing of different bulk momenta at interfaces represents a qualitative departure from the band gap engineering we have considered so far. Here the engineered changes in the band gap are accompanied by important changes in the wave function. These changes are expected to occur whenever the change in the atomic potentials across the interface is significant.

We have confined ourselves to the case where the wave mixing was due to the difference in atomic cores. This a particularly familiar case since the mechanism has been used in one of the first applications of semiconductors in optoelectronics. Perhaps some of us still remember the days when our calculators were equipped with bright green display panels (and needed strong batteries!). The light emitting diodes powering these displays were made of gallium phosphide doped with a large concentration of nitrogen impurities. The magnitude of the band gap of GaP is suitable for such an application but the optical transition is weak because the gap is indirect, i.e. the conduction band minimum lies far away from the centre of the Brillouin zone. However, when the material is doped with N, the optical emission occurs at the impurity as a recombination of an exciton bound to N (Bergh and Dean, 1975).

The nitrogen atoms replace atoms of P in the lattice of GaP. The difference between the nitrogen and phosphorous atomic potentials is very significant and the solution of the Schrödinger equation in the range of the atomic potential of nitrogen is a highly localised wave packet. If we represent this wave function following the above mentioned prescription, i.e. as a linear combination of the bulk solutions of the Schrödinger equation for perfect GaP, the sum will contain bulk waves of many different wavelengths, including a significant contribution from the bulk GaP conduction band wave function located at the centre of the Brillouin zone (at $\Gamma$). For the exciton recombination to be efficient it is necessary that the electron wave function contains a component which has a strong $\Gamma$ character. This is because the whole wave function is derived from the top of the valence band at $\Gamma$ and the optical transition is only strong when it is vertical in k-space (Bassani and Pastori-Parravicini, 1976).

Of course if we used an "ordinary" substitutional impurity such as Se to replace P in GaP the wave mixing effect would be weak and the optical transition nonexistent. This is because the difference between Se and P potentials is small and the electronic states of the Se impurity can be described by the hydrogenic model. The wave function of such a hydrogenic impurity is conceptually the same as that of a GaAs quantum well considered above. Again we have a rapidly varying part which is the bulk solution for the host material band edge with its characteristic periodicity, i.e. cos $(\pi/a)$, modulated by a slowly varying envelope function which in the ground state is of the familiar form exp $(-\alpha r)$. The Coulomb well is only a weak scatterer as is a quantum well. However, the potential difference between N and P is a very deep (Jaros, 1982) and narrow well which makes the wave function there very different from that of the bulk GaP and the simple hydrogenic envelope must be replaced by a complicated and highly localised wave packet.

The Effect of Strain Induced Momentum Mixing

So far we have assumed that it is only the difference in atomic potentials that can bring about wave mixing. However, a similar result can be obtained due to lattice mismatch (strain) (Pearsall, 1990). We can again visualise the effect upon a bulk wave function of, say, a silicon crystal, of an interface with Ge. When a thin layer of Ge is grown on a thick Si substrate, the Ge assumes the Si lattice separation in the interface plane and a larger than bulk separation in the direction perpendicular to the interface. A silicon wave function therefore sees at the interface with Ge "missing" Si atoms (vacancies) and Ge atoms sitting at the new sites in the strained lattice (Fig. 5). Hence in the range of the atomic potential of Ge the solution of the Schrödinger equation acquires a form very different from that in bulk Si (even the nodes of the wave functions lie at "wrong" positions). This wave function expressed in terms of the bulk waves of Si represents a wave packet with k-components spanning the whole width of the bulk Brillouin zone. The degree to which this packet includes waves of $\Gamma$ character (bulk conduction band waves from the k = 0 point in k space) is significant. In that case the Si-Ge superlattice can be used the same way as N was to turn an indirect gap material into a good light emitter. While both Si and Ge are indirect gap materials and consequently poor emitters of light, the Si-Ge superlattice may, thanks to the combined effect of strain and zone folding behave as a "quasi-direct" gap material.

Having explained in general terms the essence of the wave mixing effect in Si-Ge, let us now consider the case of Si-Ge superlattices in some detail. In Fig. 6 we can see the band offset diagram. Recall that the conduction band X minima and the top of the valence band of bulk cubic Ge and Si are degenerate (in fact the absolute conduction band minimum in bulk Si lies somewhat away from the Brillouin zone boundary at X (k = $\pi/a$), along the <001> direction at a symmetry point referred to as $\Delta$ (Fig. 7). The strain reduces the cubic symmetry of the bulk material and introduces an axial field described above. Hence the degeneracies at the X point

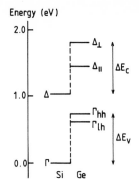

Fig. 6. The band offset diagram for Si-Ge superlattice grown on a Si substrate. hh ... heavy hole, lh ... light hole. States perpendicular and parallel (||) to the interface and showing the symmetry symbols discussed in the text are shown.

are removed and the splitting of the corresponding level can be computed as a function of the degree of strain. The larger the strain, the bigger the splitting. When Ge is grown on Si substrates, the states oriented along the <001> superlattice growth axis go up in energy and the states in the interface plane are lowered. The opposite happens when Si is grown on a Ge substrate. This is summarised in Fig. 8. The strain split levels are also shown in the offset diagram in Fig. 6; if we want to estimate the degree of confinement we must consider the barriers formed as energy differences between states in the <001> direction. In the interface plane there is no confinement.

We can begin to construct the band structure of a Si-Ge superlattice by shifting the bulk band structures of Ge and Si by the height of the band offsets including the axial splittings as

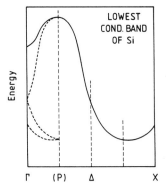

Fig. 7. A schematic picture of zone folding concerning the lowest conduction band of Si. The interrupted lines show where the bulk Brillouin zone is divided to form minibands. The minibands are shown by dotted lines folded into the superlattice or small Brillouin zone of length Γ-P.

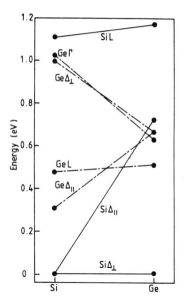

Fig.8. The relative position of the conduction $\Gamma$, $\Delta$ and L minima in Ge and Si as a function of the substrate on which the superlattice is grown. For example, the lowest point on the left corresponds to the Si conduction band minimum when the substrate is Si. On the right hand side this state is split and that of Ge remains degenerate. The bottom of the conduction band in Si is chosen always to lie at energy zero.

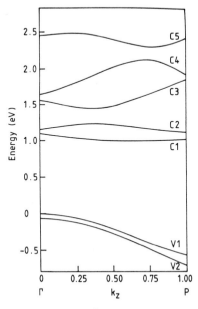

Fig.9. The band structure of Si-Ge superlattice whose period contains four monolayers of Ge and four monolayers of Si. C1 is the lowest conduction state.

indicated in Fig. 6. Since the lowest conduction states of the Si-Ge superlattice or quantum well structure must lie in Si layers, the most obvious features of the conduction band of the microstructure can be deduced from straightforward folding of bulk Si conduction bands into the small Brillouin zone of length $\Gamma$-Z (Fig. 7). By creating a superlattice of period L the bulk Brillouin zone is L times reduced and the bulk bands are folded into a small or superlattice Brillouin zone of width $2\pi/L$. This breaks the lowest conduction band of Si into L minibands.

Depending on the length of L the band minimum lies at different point in the small zone. If we wish to create a "direct" gap material we obviously desire to engineer the conduction band minimum at k = 0. This is best achieved when L is 10 monolayers wide. Then the half-length $\Gamma$-X shown in Fig. 7 of the bulk Brillouin zone is divided into five parts and the lowest point falls almost exactly at one of the dividing lines which maps onto k = 0 in the small zone.

The superlattice band structure is only approximately described in terms of straightforward application of band offsets, axial splitting and zone folding. A detailed calculation is needed to correct the energy levels for the effect of the atomic potential difference between Ge and Si and also for the details of the strain potential. As we argued above, both these effects demand that bulk states of Si and Ge mix at the interfaces so that the resulting superlattice states are in fact wave packets. This is particularly important for the wave function which we want to obtain as an expansion in terms of some bulk momentum wave functions. This then gives us an opportunity to identify the strength of the admixture into the wave function of the term representing the lowest bulk conduction state of Si at the $\Gamma$ point. We know that it is this component that must be large for the optical transition across the gap to be strong.

We can now inspect the results of large calculations which have been performed to characterise this structure in a variety of forms. For example, in Fig. 9 we see the band structure of a superlattice consisting of four monolayers of Si and four monolayers of Ge on a Si substrate. We can see that the bands obtained by zone folding in Fig. 7 are altered by the microscopic potentials. However, certain key features (i.e. the position of the minimum) are still quite recognisable.

The corresponding wave functions and optical transition strengths are also shown in Fig. 10 and Table 1, respectively.

The lowest transition happens to involve a zone folded state C2 (there is no transition from C1 for symmetry reasons) is finite because the wave function, apart from containing a strong bulk X contribution has a large bulk $\Gamma$ component. Notice that the zone folding itself merely rearranges a bulk band into a certain number of minibands and repositions the corresponding bulk wave functions. It says that because of the change in periodic boundary conditions the momentum conservation is now automatically satisfied for bulk states which map onto $\Gamma$. But it does not change the bulk energies and does not alter the bulk wave function. Hence, if we calculated the transition probability with these repositioned bulk wave functions the optical transition across the gap would still be zero. It is the effect of wave (momentum or wave vector) mixing brought about by the differences in atomic potentials and atom positions (strain) that alter the wave function from its bulk X or $\Delta$ character into a partially $\Gamma$ form. This effect is quite subtle and does not show on a wave function plot in real space in Fig. 10.

Finally, let us explore a few trends which might provide a useful guide to crystal growers and device designers. For example, in Fig. 11 we can see the change of the strength of the optical transition, which is often measured in terms of a dimensionless quantity called oscillator strength, as a function of the superlattice period. The thicker the period the weaker the optical

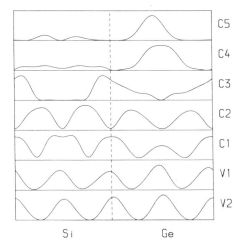

Fig. 10. The electron charge density of the key states shown in the band structure diagram Fig. 9, along the superlattice axis < 001 >.

transition! Note first of all that the hole wave function peaks in the Ge whereas the electron wave function peaks in Si layers. By increasing the period, this separation between electrons and holes increases and the optical transition probability which is proportional to the integral in space of the product of the two wave functions diminishes. But there is another more important reason for this trend. We know that in the process of the formation of band structure energies (gaps) and wave functions the Bragg constructive interference selects from the crystal potential the Fourier components which lie at the positions of the reciprocal lattice vectors. For example, in a one dimensional model it is easy to show that the magnitude of the band gap equals twice the absolute magnitude of the Fourier component of the crystal potential at multiples of $2\pi/a$ where a is the atomic separation. A superlattice is also a periodic structure and exactly the same procedure applies

Table 1. The optical transitions for the Si-Ge superlattice described in Figs 9, 10. Energies are in electronvolts. F is the oscillator strength $2mE_g | < C2 | z | V1 > |^2/\hbar^2$.

| Transition | energy (theory) | F (theory) | energy (exp) |
|---|---|---|---|
| V1 − C2 | 0.94 | 0.001 | 0.74 ± 0.14 |
| V1 − C3 | 1.37 | 0.03 | 1.25 ± 0.13 |
| V1 − C5 | 2.31 | 0.3 | 2.31 ± 0.12 |

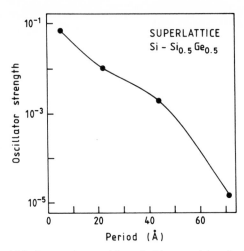

Fig. 11. The oscillator strength of the optical transition across the gap at k = 0 in a Si-Si$_{0.5}$Ge$_{0.5}$ superlattice, as a function of the superlattice period.

except that a must now be replaced with the superlattice period L. If this period is very large the only waves mixed by the superlattice potential will be those that are separated by a very small wave vector difference. However, in order to mix waves of $\Delta$ or X character which form the band minima in bulk Si with those lying at $\Gamma$ we obviously need Fourier components of the superlattice potential at a much shorter wavelength. Hence the smaller L we have the better chance there is that waves with wave vectors at $\Gamma$ (k = 0) and $\Delta$ or X will be coupled.

<u>Summary</u>

When the difference in constituent atomic potentials - or in their equilibrium positions in the lattice - across the interface exceeds a certain critical value, the bulk wave on one side of the interface cannot be simply matched to another on the other side at the same k vector. These wave functions are so different that a number of bulk waves must be combined to produce a compact wave packet which can represent this difference. This means that instead of a simple reflection expected of a rectangular barrier we must expect the wave to lose its momentum signature at the interface a couple to many waves of different bulk momenta. This process can be useful when the coupling brings into the wave function of an indirect gap semiconductor a contribution of the lowest conduction band wave function at k = 0 ($\Gamma$ point). This means that the optical transition from such a new state to the top of the valence band is partially allowed and the material acquires a direct gap character. The first example of the application of this band structure effect occurred in light emitting diodes made of N doped GaP. There the momentum mixing was due to the large difference between the atomic potentials of N and P. A similar effect is achieved in Si-Ge strained layer superlattices where the degree of admixture of the $\Gamma$ component into the conduction band wave function of the superlattice is due to the combined effect of strain, zone folding, and difference in Si and Ge potentials.

The response of a semiconductor crystal to an external photon field is normally described in terms of its refractive index n or dielectric constant $\varepsilon$. Semiconductors are highly polarisable materials with $\varepsilon$ in the static (long wavelength) limit being typically of order 10. For most purposes one assumes that $\varepsilon$ and n are independent of the intensity of the external field. We can understand the magnitude of $\varepsilon$ for a given crystal in terms of a simple semi-classical model of harmonic polarisation (Dingle, 1988). For example, we can imagine electrons as simple balls held by a spring at nuclei and oscillating with a characteristic frequency $\omega_0$. The induced dipole is a measure of the tightness with which the spring restrains the electron in its response to the polarizing field. Since the induced displacement is proportional to the applied field, this response is called linear. The dielectric constant is the induced dipole per field so that $\varepsilon$ is independent of the field.

In a quantum mechanical picture of this effect we recognise that a response of a particle means that it makes virtual transitions from its ground state to some suitable higher lying stationary level (Shen, 1984). For example, let us assume that we consider polarisation of an atomic hydrogen. Then the electron sitting initially at its ground state level, labelled in terms of the usual quantum numbers as 1s, will receive a photon and jump into the higher 2p state. Such a jump is highly probable since the corresponding optical transition probability is large. The electron will subsequently emit a photon of the same energy and jump back into the 1s level. The process is called virtual since the net energy dissipated in these two jumps up and down is nil. However, the electron wave function is on average acquiring some degree of 2p character. In fact we can write the perturbed state as a linear combination of the 1s and 2p hydrogenic wave functions. Now we know that while the 1s state is totally symmetric the 2p state is pointing in the direction of the applied field so that the original totally symmetric electron orbit has been polarised by the induced dipole in the field direction. We can of course choose to excite the system at a frequency at which the energy of the applied beam just equals the separation of the 1s and 2p levels. Then energy is absorbed and we can no longer call the process virtual. We say that the response is resonantly enhanced.

The same process can be associated with a solid except that valence electrons must be pictured in the valence band and the higher level required in the polarisation process is the conduction band. Thus the band gap replaces the energy separation between the hydrogenic (atomic) levels used in the above example. The degree of polarisation achievable in a given crystal depends on the probability with which such jumps are likely to take place.

Now suppose that a very intense beam is applied so that many more photons are available. Then it is possible to conceive not just two photon processes like the one considered above but three or four photon jumps. For an atomic hydrogen the four photon process means jumps 1s - 2p - 3d - 2p - 1s. In this process not one but two "dipoles" can be thought to have been induced and the response is a cubic function of the applied field. Hence the process is termed non-linear and, because of the cubic behaviour, a third order one. The classical analogy of this effect is an oscillator with a cubic term in the potential.

The strength of the four photon non-linear process depends not only on the availability of four photons but equally on the transition probability between the "new" high levels, in the hydrogen case in question 2p and 3d. In general the mismatch between the wave functions increases for higher states and with it the transitions between them become less and less likely. It follows that the four photon process is weak in most substances and for this reason it has never been seriously considered as a practical proposition.

When a field dependent refractive index is needed, it is customary to employ various forms of band filling (Haug, 1988). The band filling mechanism is very simple and can be explained in a common sense manner without recourse to microscopic concepts. A holding beam is used to excite electrons from the ground state to an upper level. The probe beam operated at a slightly lower frequency is then unable to find suitable empty states into which to deposit electrons excited from the ground state. Hence the system as seen by the probe beam appears to have a very different refractive index. This way a very large change in the refractive index can be achieved. Unfortunately, the effect is accompanied by a large amount of energy dissipation which makes the process difficult to implement in devices. Hence the search for new materials exhibiting large optical non-linear response at reasonable fields is still on.

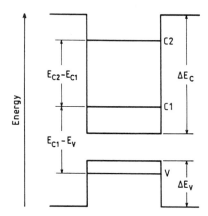

Fig. 12. A level diagram describing band offsets and level separations required to achieve optical non-linearity.

The advent of molecular beam epitaxy opened the way for engineering novel electronic structures and optical spectra. In particular, we shall see that it is possible to design semiconductor superlattices in which the non-linear response is exceptionally large. Most estimates of the magnitude of the third order susceptibility $\chi^{(3)}$ lead to an expression of the form

$$\chi^{(3)}(\omega,-\omega,\omega) = -e^4 p_{01}p_{12}p_{21}p_{10}/$$
$$\{V \hbar^3 m^4 \omega^4 (\omega_{10}-\omega)(\omega_{20}-2\omega)(\omega_{10}-\omega)\} \tag{1}$$

where $\omega_{ij} = (E_i - E_j)/\hbar$ and $p_{ij} = <i \mid z \mid j>$ are the dipole matrix elements, in the usual bra and ket notation, representing the probability of an optical jump between states $i$ and $j$. This $\chi^{(3)}$ will be very large when the dipole transitions are all allowed and when at the same time the frequency differences in the denominator nearly vanish. Note that we have not included the damping constants

160

(relaxation times) which would make energies complex and ensure that the denominator does not become exactly zero at resonance. Such a more detailed parametrisation is necessary when it comes to evaluation of device performance etc. but it does not interfere with the basic idea which is what we want to get across here.

This condition for the implementation of the idea is pictured in Fig. 12. The figure also indicates how such a situation might be created in a semiconductor quantum well system. State 0 can be identified as the uppermost valence state, states 1 and 2 are the lowest two conduction states. These two states must be separated by an energy difference approximately equal to the principal gap. This gap must occur at the centre of the Brillouin zone which means that the constituent bulk materials in question must be direct gap crystals. Only then can we be sure that the integrals $< i \mid z \mid j >$ are large. The two constituent crystals must also be well lattice matched because we shall require as little perturbation at the interface as possible.

Let us now consider some suitable candidates. In general, we are looking for a system where the conduction band offset is so large that the separation between the lowest two confined conduction levels is comparable to the band gap. The valence band offset must be much smaller but such that both the electron and hole states are confined in the same layer. One such candidate is a quantum well structure of InAs and ZnTe. The band gap of InAs is small (0.42 eV) and the conduction band offset with ZnTe large (1.27 eV). The valence band offset is about 0.7 eV. Ignoring the small lattice mismatch (0.8%), it transpires that the required energy separations can be realised, for example, in a structure whose period consists 21 monolayers of InAs (i.e. 63.5 Å) and 13 monolayers of ZnTe (39.6 Å). The separation is then about 0.635 eV. Another example is $CdTe-Hg_xCd_{1-x}Te$. The superlattice comprising 10 monolayers of CdTe (26 Å) and 8 monolayers of $Hg_{0.82}Cd_{0.18}Te$ (32 Å) exhibits a band gap of about 0.39 eV.

Table 2

A ... barrier (AlSb) width in angstroems B...well width in angstroems x determines the percentage of GaSb in the alloy of InAs and GaSb forming the well material. $E_g$ is the band gap in electronvolts. The parameters A, B and x were chosen so as to satisfy the condition for large optical non-linearity.

| x | A | B | $E_g$ (eV) |
|---|---|---|---|
| 0.88 | 30.4 | 30.4 | 1.103 |
| 0.54 | 30.4 | 42.6 | 0.884 |
| 0.23 | 30.4 | 54.7 | 0.725 |

Finally, let us consider a group of III-V well lattice matched semiconductors InAs, GaSb and AlSb. Their band offsets are summarised in Fig. 13. We can see that here again the conduction band offset between AlSb and the other two materials is large enough to allow for two well confined levels separated by an energy difference comparable with the forbidden gap. Although AlSb itself is an indirect gap material, the position of the secondary minima of X character is so high in energy that it is unlikely to play a significant part in determining optical spectra. We can therefore ignore the indirect gaps. By choosing AlSb as barrier material, we are free to pick an alloy of InAs and GaSb and the well width so as

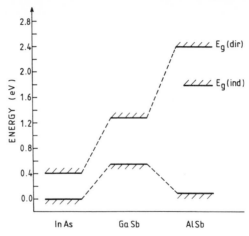

Fig. 13. A band lineup diagram for InAs, GaSb and AlSb. dir and ind
indicate direct and indirect (at the X point) gaps of the
bulk crystal.

to achieve a particular operating frequency ω (i.e. the energy
separation of the upper levels and the gap). In Table 2 we give a
few examples of structures satisfying the above mentioned criteria.
The choice of the barrier width given in the table is immaterial and
the system works as a multiple quantum well structure. Naturally,
when the barrier width is reduced, a superlattice is formed and the
energy levels change accordingly. We can see that this choice of
material spans the technologically important 1.55 μm range of
wavelengths.

While the examples given above clearly illustrate the
underlying physical principle, and do lie within the current MBE
capabilities, both the HgTe-CdTe and InAs-ZnTe structures are among
the most demanding materials. In HgTe-CdTe the interdiffusion makes
it difficult to characterise the interface quality and the well
width is likely to fluctuate from layer to layer. The polar
interfaces of InAs-ZnTe are not yet well understood. Not only is Zn
and Te readily diffusing into the other constituent. The large
difference in ionicity across the interface is likely to lead to
defects and irregularities. The growth technology of AlSb is also
relatively poorly understood. However, it is important to realise
that the structure design described above does not impose very
strict requirement on the well width fluctuations or on interface
quality. Since we do not expect the device to be operated at
resonance, the details of the inhomogenous line broadening are not
essential. The proposed structures therefore represent a fresh and
realistic step towards designing novel nonlinear materials for
future experimentation in non-linear optics and for needs in
optoelectronics.

References

Bassani, F. and Pastori-Parravicini, G. (1975) *Electronic States
    and Optical Transitions in Solids* (Pergamon, New York).
Bergh, A.A. and Dean, P.J. *Light Emitting Diodes* (1976) (Oxford
    University Press, Oxford).

162

Capasso, F. and Margaritondo, G. (Eds) (1987) *Heterojunction Band Discontinuities: Physics and Device Applications* (North Holland, Amsterdam).

Dingle, R. (Ed.) (1988) *Applications of Multiquantum Wells, Selective Doping and Superlattices*, Semiconductors and Semimetals Vol. 24 (Academic, New York).

Haug, H. *Optical Non-linearities and Instabilities in Semiconductors* (1988) (Academic, London).

Jaros, M. *Deep Levels in Semiconductors* (1982) (Hilger, Bristol).

Jaros, M. *Physics and Applications of Semiconductor Microstructures* (1989) (Oxford University Press, Oxford).

Pearsall, T.P. (Ed.) *GaInAsP Alloy Semiconductors* (1982) (Wiley, New York).

Pearsall, T.P. (Ed.) *Strained Layer Superlattices*, 1990, Semiconductors and Semimetals (Academic, New York).

Shen, Y.R. *The Principles of Non-Linear Optics* (1984) (Wiley, New York).

# BASIC OPTICAL PROPERTIES OF LOW DIMENSIONAL STRUCTURES

# FOR APPLICATIONS TO LASERS, ELECTRO-OPTIC AND

# NON-LINEAR OPTICAL DEVICES

C. Weisbuch

Thompson-CSF, Laboratoire Central de Recherches
91404 Orsay, France

The purpose of this set of lectures is to introduce students to the field of the optical properties of Low Dimensional Structures (LDS's) and their applications. As this field has already reached immense dimensions and is still growing very fast, we only cover some very fundamental items, most of them of universal use in LDS's, with illustrations only in the direct-gap type I quantum well, examplified by the GaAs/GaAlAs system. We have chosen to cover the following few topics :

-Basic optical properties of LDS's, and their relation to optical properties of isolated atoms and 3D solids.

- Selected illustrations of the linear optical properties of quantum wells.

- Quantum well lasers

- Electro-optical and non-linear optical properties of Quantum wells

- Fundamentals and applications of quantum well wires (QWW) and dots (QD).

This set of lectures therefore does not discuss many topics concerning the optical properties of quantum wells, which have been covered in other lectures at this school or can be found in recent reviews : a wide set of topics has been selected for publication in a special issue of the Journal of Luminescence, and is beautiffully introduced by Sturge and Meynadier[1]. Specific references are calculation of energy levels[2,3,4] optical properties of type II quantum wells and superlattices[2], $Ge_x Si_{1-x}$ QW's and superlattices[5,6], II-V QW's and superlattices (both wide[7] or narrow gap[8]), strained-layer QW's and superlattices[9], light-scattering phenomena[10], optical hot-carrier phenomena as studied by optical techniques[13],etc... Time resolved spectroscopy will only be mentioned briefly in direct relation with parameter determination and applications, but we refer to recent excellent reviews of this fast expanding field[14].

## II - BASIC OPTICAL PROPERTIES OF LDS's

1. Basic optical properties of atoms and solids

It is instructive to first detail a simple theory of the optical properties of atoms and solids and compare these. As should become clear in these lectures, going to LDS's yields better optical properties because one transforms the band-broadened optical parameters of solids to sharp, atom-like resonances.

Atoms lead to very simple descriptions of optical properties of solids. The classical Lorentz model [15] of an atom as damped, elastically-bound charge leads to the classical equation of motion along x

$$m \frac{d^2x}{dt^2} + m\gamma \frac{dx}{dt} + kx = F \tag{1}$$

yielding a forced motion, under excitation by a x-linearly-polarized plane wave $E = E_o \exp[i(\omega t - kz)]$, given by (for electrons with -e charge)

$$x = \frac{-e\,E_o}{m} \frac{\exp i(\omega t - kz)}{\omega_o^2 - \omega^2 + i\omega\gamma} \tag{2}$$

where m is the free electron mass, $\gamma$ is the damping constant, k is the stiffness of ionic binding, $\omega_o$ is the resonance pulsation ($\omega_o = (k/m)^{1/2}$). For volume density of N atoms, (2) leads to a volume electric polarization P

$$P = \sum -ex = \frac{Ne^2\,E}{m} \frac{1}{\omega_o^2 - \omega^2 + i\omega\gamma} \tag{3}$$

which defines the linear electronic susceptibility $\chi^1(\omega)$

$$\chi^1(\omega) = \frac{Ne^2 / \varepsilon_o\,m}{\omega_o^2 - \omega^2 + i\,\omega\gamma} \qquad \text{with } P = \varepsilon o\,\chi^1\,E \tag{4}$$

From there, one can define the propagation equation using Maxwell equations and the electric displacement vector D defined by

$$D = \varepsilon_o\,\varepsilon_b\,E + P = \varepsilon_o\,\varepsilon_b \left(1 + \chi^1 / \varepsilon_b\right) E = \varepsilon_{tot}\,E$$

where $\varepsilon_b$ is the background permittivity due to all causes other than the atom transition under study, and $\varepsilon tot$ is the total permittivity at frequency w. One usually defines additional interrelated quantities such as the complex permittivity $\varepsilon_{tot} = k_{1-} ik_2$, complex index of refraction $n_{tot} = \sqrt{\varepsilon_{tot}} = n_r - i\,n_i \sim n_b (1 + \chi^1 / 2\varepsilon_b)$, complex susceptibiliy $\chi 1 = \chi' - i\,\chi''$. When such quantities are injected in the propagation equation, they yield propagating waves with an intensity attenuation factor $\alpha = 2n_i\,k_o = n_b\,k_o\,\chi''(\omega) / \varepsilon_b = k\,\chi''(\omega) / \varepsilon_b$ and dephasing due to the considered transition given by $n \cong \chi'(\omega) / 2\varepsilon_b$. Here $k_0$ is the vacuum propagation wavevector and $k = n_b k_o$ is the medium propagation wavevector with background index $n_b$.

As is well-known, the real and imaginary parts of $\varepsilon tot$, $n_{tot}$, $\chi 1$ are related through the Kramers-Kronig relation, which means that the experimental determination of one quantity (such as the absorption coefficient) allows the calculation of the index of refraction.

The quantum mechanical calculation af atomic susceptibilities yields, for a perturbative two-level model [16,17],

$$\chi'(\omega) = \frac{N}{\varepsilon_o\hbar} |\langle 1 | e\vec{r} | 2 \rangle|^2 \frac{\omega_o - \omega}{(\omega_o - \omega)^2 + \gamma^2 / 4} \tag{5}$$

$$\chi''(\omega) = \frac{N}{\varepsilon_o\hbar} |\langle 1 | e\vec{r} | 2 \rangle|^2 \frac{\gamma / 2}{(\omega_o - \omega)^2 + \gamma^2 / 4} \tag{6}$$

where $\langle 1|e\vec{r}|2\rangle$ is the electric dipole operator matrix element and $\gamma$ is the FWHM of the transition. Comparison with expression (4) **near a resonance** yields an identity provided that one replaces the "classical" squared charge $e^2$ by $e^2f$, where f is the **oscillator strength** given by

$$f = 2\, m\omega\hbar^{-1}\,|\langle 1|\vec{r}|\, 2\rangle|^2 = 2\,(m\omega\hbar)^{-1}\,|\langle\, 1|\vec{p}|2\,\rangle|^2 \qquad (7)$$

From (5), (6) and (7) one infers that many physical quantitites are directly related to f. For instance, considering an isolated resonant transition, one can deduce that the energy-integrated absorption coefficient is :

$$\int_{\text{absorption line}} \alpha\,(\omega)\,d\omega = \frac{\pi e^2}{m\,\varepsilon_o\,n_b}\, Nf \qquad (8)$$

Figure 1 (top) shows the measured values for $n_r$ and $n_i$ in GaAs. One remarks that although the general features of (5) and (6) are reproduced for two resonances, one around 36 meV for ionic motion, the other around 2-3 eV for valence electronic transitions, the resonances are extremely broadened compared to an atomic resonance, which usually has $\gamma \approx 10^7 s^{-1}$. This is the more remarkable for the behaviour around the GaAs bandedge at 1.42eV, (figure 1 bottom) where instead of a sharp dispersion-like curve for $n_r$ one only observes a small cusp when $\hbar\omega$ reaches the gap energy. The mathematical expression of the cusp can be obtained by the Kamers-Kronig transformation of the usual absorption edge energy dependence $\alpha \sim \sqrt{E - E_g}$ .This weak effect of the absorption edge is of course due to the spreading of the transition energies and wavectors of the N valence electrons over energy band states .

The calculation of $\chi^1$ in solids is exactly given by expressions (4) or (5) and (6), implementing the correct oscillator strength and summing over all possible transitions at various frequencies $\omega_o$.

**As will be seen below, one of the main effect of LDS's will be to transform such band-to-band broad transitions to sharp atom-like transitions through excitonic effects and/or k-matching through quantum-confinement.**

## 2. Oscillator strengths

The optical matrix element entering the oscillator strength in a solid usually retains an atomic value[18]. This is due to the delocalized nature of the Bloch wavefunction in a solid. We will demonstrate it here in the case of **interband transitions in a quantum well** as the extension towards 3D or 1D and 0D is straightforward. The electron (or hole) wavefunction is given by (in the enveloppe wavefunction approximation)

$$\Psi_{e(h)}(\vec{r}) = \frac{1}{\sqrt{V}}\, e^{i\vec{k}_{e(h)\perp}\cdot\,\vec{r}_\perp} u_{c\vec{k}_{e(h)}}(\vec{r})$$

For a quantum well, the interband optical matrix element has the form (neglecting the photon wavevector)

$$M = \langle f|\vec{r}.\vec{\eta}|\,i\rangle = \frac{1}{V}\int_{\text{QW volume}} \chi_e(z)e^{i\vec{k}_{e\perp}\cdot\,\vec{r}_\perp}\, u_{ck_e}(\vec{r})\vec{\eta}.\vec{r}\, \chi_h(z)e^{i\vec{k}_{h\perp}\cdot\,\vec{r}_\perp} u_{vk_h}(\vec{r})\, d\vec{r} \qquad (9)$$

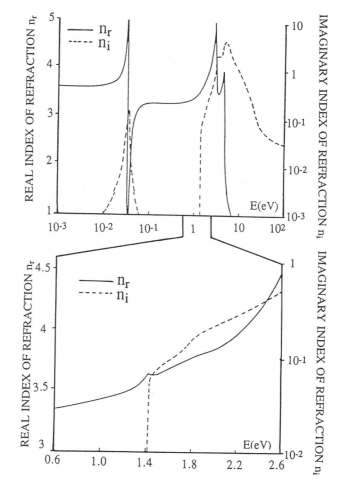

Figure 1 . Real ($n_r$) and imaginary ($n_i$) index of refraction for GaAs (compilation courtesy of R. Baets)

where V is the quantum well volume, $\chi_e(z)$ and $\chi_h(z)$ are the electron and hole envelope wave functions, $\vec{k}_e$, $\vec{k}_h$ are electron and hole wave vectors, $\vec{\eta}$ is the polarization vector of light, $u_{ck_e}(\vec{r})$ and $u_{vk_h}(\vec{r})$ are the usual Bloch functions. The integral contains fast-varying functions over unit cells ($u_{ck}$ and $u_{vk}$) and slowly varying functions (figure 2a). Using the usual procedure, one transforms Eq.(9) in a summation of localized integrals involving only Bloch functions over the N crystal unit cells labelled by their centers $\vec{R}_i$ :

$$M \sim \frac{1}{V} \sum_{R_i} \chi_e(\vec{R}_i)\,\chi_n(\vec{R}_i)\,e^{i\left(\vec{k}_{h\perp} - \vec{k}_{e\perp}\right)\cdot\vec{R}_i} \int_{cell} u_{ck_e}(\vec{r})\,\vec{\eta}\cdot\vec{r}\,u_{vk_h}(\vec{r})\,d\vec{r} \quad (10)$$

The latter integral is independent of $\vec{R}_i$ and is $\Omega P$ where $\Omega$ is the unit cell volume and P the usual interband matrix element (independant of dimensionality of crystal) which contains the selection rules due to the band symmetries and light polarization. When summing for the transverse directions, the exponential oscillatory factor gives a null contribution unless $\vec{k}_{e\perp} = \vec{k}_{h\perp}$, which is the vertical transition selection rule (if we had kept the photon wavevector $k_{ph}$ in (9) we would obtain at this point $\vec{k}_{e\perp} = \vec{k}_{h\perp} + \vec{k}_{ph}$, the wavector conservation relation).

The $R_i$ transverse summation then yields the quantum well surface. The only difference between equation (10) and the usual 3D summation lies in the z-direction summation, which produces a factor $\sum \chi_e\left(\vec{R_i}\right) \chi_h\left(\vec{R_i}\right) a$ , where the $R_i$'s are the lattice cell centers in the z direction and a is the lattice constant. Transforming back into an integral $\quad \chi_e(z) \chi_h(z) dz$, one finds a unity factor for the transitions between electron and hole states with the same quantum number n, as the $\chi$'s are identical $[\sim \sin(n\pi z/l)]$ and normalized to unity. **Therefore the interband matrix element has an atomic value P, and this result can be shown to be independant of crystal volume and dimensionality.** In any system, the oscillator

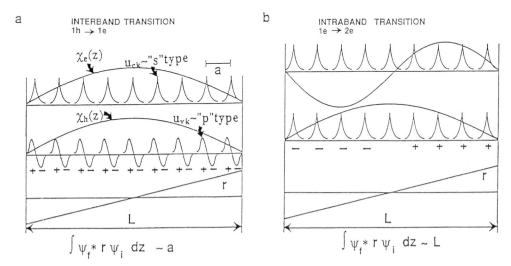

Figure 2 . (a)Schematics of the integrand function of the interband matrix element as the product of the functions, i.e. $\chi_e(z)$, $\chi_h(z)$, $u_{ck}$, $u_{vk}$ and r. Note the inversion of $u_{vk}$ at each lattice site, which determines equal contributions to the total dipole moment. (b) Same for interband transitions in the conduction. Sites which are quite away from the well center and extremity contribute more to the dipole moment, which leads to the giant dipole effect.

strength per unit volume is thus the energy density of quantum states (more precisely transitions) per unit volume times the atomic oscillator strength. The 2D confinement and lower dimensionality of QW's therefore does not change the oscillator strength per electron-hole pair. However, when comparing 3D and 2D systems, as is clear from the density-of states curves (figure 3a, b), quantum confinement leads to a better k-matching of electron and hole wavefunctions : in the z-direction all electrons and holes in a same quantized subband have same $k_z$ wavevector which "concentrates" the oscillator strength when compared to the 3D case. This point becomes more evident and important when going to still lower dimensions (figure 3c, d). In the absence of exciton effects in quantum wells, the absorption coefficient should reflect the reduced 2D DOS, i.e., should consist of square steps corresponding to the various confined states.

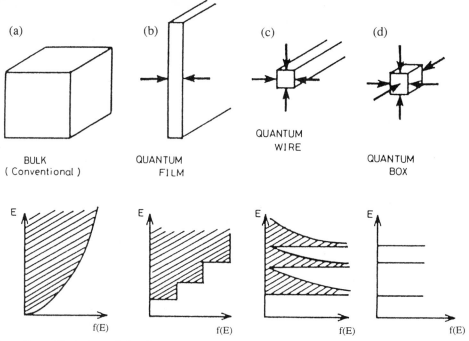

Figure 3 . Variation of density of states, i.e. of spectral oscillator strength, with varying dimensionality of heterostructures. (after Asada et al.[19])

The matrix element M given in equation (9) takes a much different shape when dealing with intersubband transitions, i.e. between confined electron states or hole states only (figure 2b). In that case, the fast-varying integrals involve the same periodic part of the wavefunction, which have then zero matrix element with $\vec{r}$ . Equation (9) becomes

$$M \sim \frac{1}{V} \int_{crystal} \chi_e(z) \, \vec{\eta} \cdot \vec{r} \, \chi'_e(z) \, d \, r \int_{cell} u_{cke}(\vec{r}) \, u^*_{cke}(\vec{r}) \, d^3 r \qquad (11)$$

when using the usual procedure of transforming the M-integral into a summation over fast varying contributions and transforming back into an integral. The second integral yields unity when using normalized Bloch functions. The matrix element has now large values, of the order of L, dimension of the quantum well, instead of an atomic dimension, like in interband transitions. This "giant" dipole effect was first observed by West and Eglash[20]. The value of the dipole moment between the ground state and first excited state is, in the infinite well approximation,

$$e <z> = (16 / 9\pi^2) eL \qquad (12)$$

Applications of the giant intersubband dipole are already numerous. Levine and his team[21] have developed a number of detectors on this basis. This detection scheme seems to be intrinsically inferior to interband-based detectors such as HgCdTe alloys due to the very short lifetime in the excited state ($\sim 10^{-12}$s through L0 phonon emission instead of $10^{-9}$s through recombination) but could still be of great usefulness due to the much better materials properties of GaAs and monolithic integration of detector and electronic functions.

## 3. Excitonic effects

As is well-known[22], a photocreated conduction electron interacts via the Coulomb interaction with the hole left behind in the valence band to yield an hydrogen-like complex, the exciton.

Besides providing an energy shift of the absorption band to well-defined exciton absorption peaks, the electron-hole correlation concentrates the light-matter oscillator strength. This is most readily seen by examining the relative motion of the electron and hole in the exciton (figure 4) : clearly, electron and hole wavefunction overlap much more strongly than in the case of delocalized free electron-hole pairs. The oscillator strength **per crystal unit volume** can be shown to be **in 3D**, after a lengthy but straightforward calculation[22]

$$f = \frac{2m\omega \, \left| \left\langle \psi_c \left| \vec{\eta} \cdot \vec{r} \right| \psi_v \right\rangle \right|^2}{\hbar} \cdot \frac{1}{\pi a_B^3} = f_{at} \cdot \frac{1}{\pi a_B^3} \tag{13}$$

where $\left\langle \Psi_c \left| \vec{\eta} \cdot \vec{r} \right| \Psi_v \right\rangle$ is the standard valence band to conduction band matrix element and $a_B$ the Bohr radius. This value can be qualitatively justified from close examination of the exciton wavefunction (figure 4). It also represents the number of excitons which can be closely packed in a unit volume crystal. Therefore, if the exciton line is narrower than the Rydberg energy, oscillator strength has been concentrated at the exciton line energy by the electron-hole interaction.

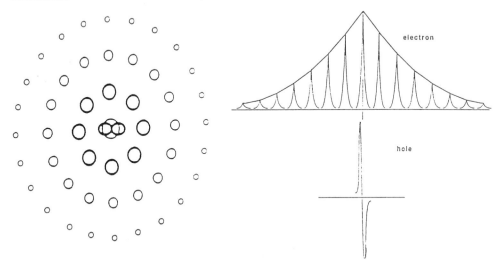

Figure 4 .     Schematics of the 2D extension [101] and 1D amplitude of the relative exciton wavefunction. The hole has been considered much heavier than the electron, hence its relative localization.

In 2-D, a similar concentration of oscillator strength from free states into the exciton state also exists. The infinite well approximation yields, in the exact 2D limit, an oscillator strength per unit surface[23]

$$f = \frac{2m\omega\left|\left\langle u_C \left| \vec{\eta} \cdot \vec{r} \right| u_V \right\rangle\right|^2}{\hbar} \cdot \frac{8}{\pi a_B^2} = f_{at} \cdot \frac{8}{\pi a_B^2} \qquad (14)$$

where use of the 4 times smaller 2D Bohr radius when compared to 3D has been made, $a_B$ retaining here its 3D value.

One therefore expects two main effects from the diminition of dimensionality when comparing 3D and 2D excitons :

- The **increase in binding energy** of excitons in 2D (up to 4 times the 3D Value) leads to resolved exciton peaks at room temperature, as the broadening due to exciton ionization through LO-phonon collision is smaller than the measured Rydberg (1.5 meV vs 9 meV)[23, 24].

- The **concentrated oscillator strength** is larger than in 3D, due to the reduced Bohr radius. The integrated absorption peak is therefore ~16 times larger in the exact 2D limit, never reached as evidenced by the more precise calculations and measurements of the exciton binding energy (9 meV instead of 16 meV). However, as seen on figure 5, the observed absorption peak at room temperature depicts a large resonant enhancement of the light-matter interaction which is used in many applications.

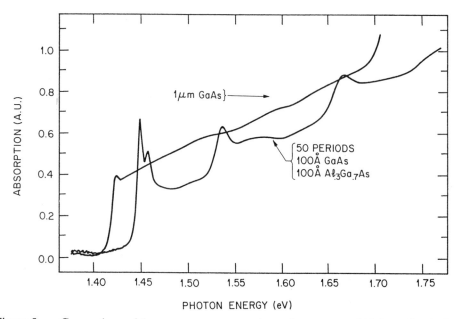

Figure 5 .   Comparison of the room-temperature absorption spectra of high-quality bulk and $L_z = 100\text{Å}$ MQW GaAs samples. The bump at the onset of bulk absorption is the remnant of the thermally-dissociated exciton. The sharp exciton peak in MQW's denotes a stronger resonant light-matter coupling[24].

The increase in light-matter coupling due to 2D excitons has been studied in details in the CdTe-Cd ZnTe case[25]. Figure 6 shows the variation of the integrated transmittance (integrated absorption strength per unit area) evidencing the 2D character of excitons below 200Å, with a equivalent volume f corresponding to $\approx 5.10^{19}$ cm$^{-3}$ CdTe molecules (along equation 8).

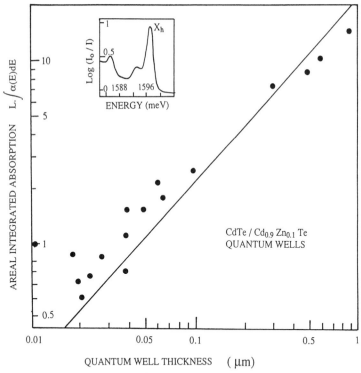

Figure 6 .    Integrated transmittance  (L . α(E) dE : proportional to areal oscillator strength)
for CdTe quantum wells. Insert shows a typical exciton absorption spectrum


4. Selection rules

    Zero-order selection rules on the optical matrix element M and radiation patterns can
easily be deduced from atomic-physics analogies as it is known from the correspondance
principle that quantum -mechanical electric dipoles (determined by the matrix elements of the
position operator $\vec{r}$ ) absorb and radiate like classical oscillating dipole moments (in particular :
emitting or absorbing dipole and absorbed or radiated light field have same polarization vector ;
emission or absorption pattern goes as $\sin^2 \theta$ where $\theta$ is the angle between the dipole  axis and
the light wavevector) . One can then deduce the absorption and radiation pattern for a transition
between two quantum states, provided one calculates the electric-dipole matrix elements
between these two states. Figure 7a shows the dipole moment between an electron s-state and a
hole p-state, and 7b represents the resulting radiation pattern. If one neglects band-mixing in
quantum wells at zero wavevector, the dipole moments are shown in fig. 7c. The emission
diagram is simple to calculate and is shown in fig. 7d. This description is however somewhat
oversimplified as in a solid one has to sum possible transitions over all k-directions, but  this
summation yields the results given in the atomic picture of fig. 7c [18]. Remembering that a
classical linear electric dipole radiates light which is linearly polarized along the motion axis of
the electron, one can deduce for instance that both heavy and light holes are coupled to photons
in the TE mode propagating along the quantum well plane, whereas only light holes are coupled
to TM polarized light.

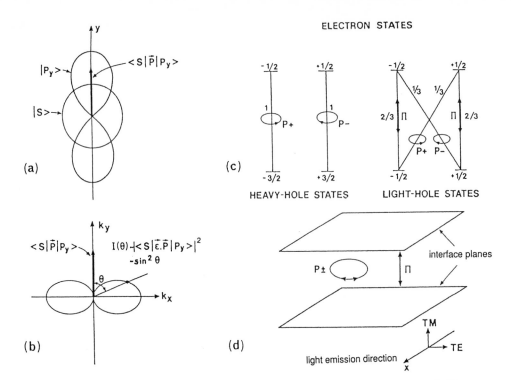

Figure 7 .    Optical selections rules for absorption and luminescence between atomic-like states
(Bloch States) of valence and conduction bands

In semiconductor QW's, for transitions occuring with non-zero transverse wavector, the anisotropic heavy-hole to electron matrix element introduces a gradual change of selection rules with kinetic energy. Asada et al.[26] have shown  that, using the $\vec{k}.\vec{p}$ perturbation theory to calculate electron and hole wavefunctions at $k_\perp \neq 0$, a rather simple geometric calculation of the interband matrix element between electron and heavy holes could be made. The dipole moment is rotating in the plane perpendicular to the total k electron wavevector $\vec{k} = \vec{k_z} + \vec{k_\perp}$. For a given  energy $E = E_0 + E_{conf} + \hbar^2 k_\perp^2 / 2m$, the locus of positions of $\vec{k}$ is a constant $|\vec{k_\perp}|$ circle around $k_z$. The averaged value of $\vec{\eta}.\vec{r}$ can then easily be determined for given transition.

At zero order, excitons tend to retain these selection rules as long as band mixing effects are neglected. This is usually done as the main features observed in polarized luminescence experiments reproduce qualitatively the expected polarization behaviour. Such experiments proved to be extremely useful to label luminescence peaks[27,28].

An important set of selection rules originates in the overlap integral $< \mathcal{X}_e \mid \mathcal{X}_h >$ originating from eq. (10). In the infinite well approximation, $\mathcal{X}_e$ and $\mathcal{X}_h$ are usually orthogonal, i.e. their integral is zero unless they have equal quantum numbers, hence the $\Delta n = o$ selection rule of Dingle. For finite quantum wells, the $\chi$'s are no more exactly orthogonal as the penetration of the wavefunction in the barrier material depends on the quantum number.

However, the single-particle wavefunctions retain they parity (even or odd), due to the symetric confinement potential and one therefore only expects weak $\Delta n \neq 0$ transitions with both n's being either even or odd. When the single particle picture breaks down, such as at high densities, other symmetry-breaking transitions can be observed. They become of course of paramount inportance when a symmetry-breaking perturbation is applied, such a transverse electric field.

## III - SOME RESULTS OF OPTICAL SPECTROSCOPY

The main optical techniques are photoluminescence, absorption, excitation spectroscopy, reflectivity and modulation spectroscopy. The various qualities of these techniques have been previously described[28] and are also discussed in Professor Hamilton lectures at this school[29]. We will therefore only illustrate here some of the recent results on intrinsic QW properties.

Whereas the bulk of attention has been focussed on the size quantization of energy in QWs light emission, the most evident and ubiquitous effect is the much **larger quantum efficiency** of photoluminescence than for 3D bulk material in reported in all materials systems studied up to now : GaAs/GaAlAs, GaInAs/InAlAs, GaSb/GaAlSb, GaInAs/InP, CdTe/CdMnTe, GaAsSb/GaAlSb, ZnSe/ZnMnSe etc... The origin of this enhanced radiative efficiency is not quantitatively worked out and might be the result of several effects, on radiative or non-radiative recombination, depending on the materials systems and on growth conditions :

- due to the larger exciton binding energy than 3D, correlation effects increasing the electon-hole overlap for efficient light emission can be stronger and exist at higher temperatures.

- as evidenced by Petroff et al.[30] interfaces can act as efficient getters for impurities which would otherwise act as non-radiative centers. The smoothing action of interfaces also diminishes structural defects and therefore non-radiative recombination centers. The improvement in quantum-efficiency of material grown after a superlattice buffer layer has been widely reported.

- dislocations in QWs have been shown to be inactive as non-radiative centers, although selective etching reveals that dislocations are present[31]. No complete explanation of this phenomenon has been given which could either due to inefficient capture by dislocations in 2D or by diminished gettering of impurities by dislocations due to the more efficient gettering by interfaces.

- lower activity of impurities : extrinsic recombination processes involving impurities (electron to neutral acceptor recombination, bound excitons, etc...) only appear in QWs at significantly higher impurity concentrations than in 3D. This might be due to the more efficient intrinsic radiative recombination mechanisms or to inefficient 2D impurity capture mechanism, although this last point has not been theoretically evaluated.

Besides providing very important information on the physics of quantum wells, <u>perturbed luminescence</u> brings detailed information on the mechanisms through which quantum well devices will operate. The large effects of applied electric fields and of carrier band filling effects are evidenced both in cw and transient experiments. For interband transitions, it is well-known that modifications of band edge optical parameters will come from either one or several of the following effects[24] :

- energy shift of the bandedge due to an applied electric field (Franz-Keldysh effect), renormalization due to carrier- induced many-body effects or carrier bandfilling ("Burstein-shift")

- screening of interactions such as the carrier-induced Coulomb interaction screening of excitons

- dielectric response of carriers (plasma optical response)

- spatial charge transfer under electric fields, within homogeneous regions or across regions with spatially-varying doping or composition, such as in tunneling structures. The control of second-order effects such as carrier lifetime increase is essential when electron and hole are spatially separated under an applied electric field[31].

The detailed review of these effects is way beyond the scope of these lectures[24]. For our purpose, the experiments displayed on figures 8-9 are sufficient to illustrate the effects which will be used in devices. The Quantum-Confined Stark Effect[24] (figure 8) yields energy shifts which are much larger than the 3D Franz-Keldysh effect both due to the fact that energy levels can be strongly modified by applying large electric fields while still retaining a good oscillator strength as the electron-hole overlap remains finite due to the confinement of the electron and hole against either interface of their common quantum well. An additional bonus of this confined situation is the remain of some excitonic correlation effect in the oscillator strength even at the largest applied electric field whereas in 3D the exciton is discociated at very moderate fields.

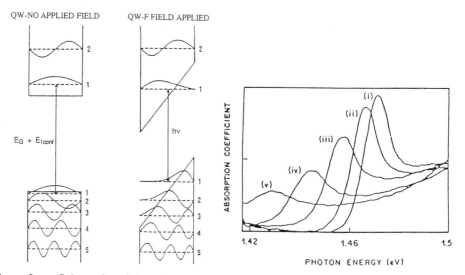

Figure 8 .    Schematics of the action of an electric field on a quantum well (Quantum confined Stark Effect) and measurements on a 94Å MQW sample for various fields (0,6.10⁴, 1.1 10⁵, 1.5 10⁵, 2.10⁵ V. cm⁻¹ respectively)[24].

Carrier-induced effects on the absorption edge are evidenced in figures 9, either through gate-control[32] (9a) or light excitation[33] (9b). In both cases the dominant change in the absorption edge spectrum is the disappearance of the strong excitonic features at very moderate densities ($\approx 5.10^{11}$ cm⁻²). This is due to the exciton screening by free carriers at such densities, when the mean distance between carriers is such that on average a carrier is present within an exciton volume, which destroys the Coulomb correlation between electron and hole leading to exciton formation. As can be seen, band filling effects occur at large densities and only lead to smooth modifications of the absorption edge. Such features are conserved at room temperature, due to the existence of exciton absorption peaks in QW's, a unique feature among semiconductors.

The above-described modifications of the absorption edge will of course lead to important variations of the refractive index, as can be calculated through Kramers-Kronig analysis.

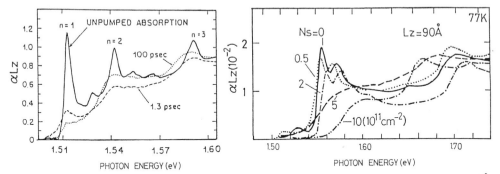

Figure 9 . Absorption changes in MQW's (77K) due to photocreated carriers[32] (left : 250Å QWs, exciton density 5.10[11]cm[-2], absorption observed 1.3 or 100 ps after exciton in a picosecond pump-probe experiment) or field-induced carriers[33] (right : 90Å QWs, carrier density changed in a filed-effect transistor)

## III - QUANTUM-WELL LASERS

### 1. Background on semiconductor lasers[17]

i) Lasers are based on self-oscillations of light emitting systems through stimulated emission. In order to have gain overcome absorption, one needs to have inversion of the quantum states between which the laser transition is to occur. From inspection of the shape of the Fermi-Dirac distribution function, this means that one has to fill the band states of a semiconductor laser at least up to an energy $\approx kT$.

ii) One uses emission in an optical cavity in order to concentrate the emitted photons in only a few modes through reflections back and forth on the mirrors. As the quantum stimulated emission probability is proportional to the number of photons per mode, one increases in that way the efficiency of emitted photons.

iii) A main parameter is the optical confinement factor $\Gamma$ which represents the fraction of the optical wave in the active layer material, i.e. the efficiency of an emitted photon to interact with another e-h pair in order to further induce stimulated emission. This confinement factor depends on the active layer thickness and on the difference in index of refraction between the active layer and confining materials. For narrow wells below 500Å, $\Gamma$ is well approximated by

$$\Gamma \approx k_0^2 \left(n_2^2 - n_1^2\right) d^2 / 2 \tag{15}$$

$n_1$ and $n_2$ being the indices of refraction of the well and confining layer materials respectively.

As indicated in figure 10, the $d^2$ variation of $\Gamma$ for simple heterostructures (because of diminished overlap of an optical wave which becomes wider with decreased active layer material) leads to extremely small values of $\Gamma$ in quantum wells, typically $4.10^{-3}$ for $d = 100$Å. Using the separate confinement heterostructure (SCH) scheme, with a fixed cavity to confine the optical wave separately from the electron wave (quantum well), one obtains $\Gamma \sim 3.10^{-2}$ in a 100Å GaAs/GaAlAs QW.

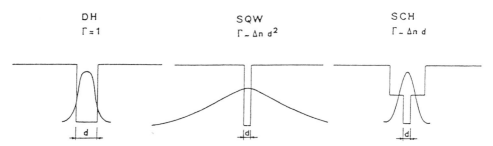

Figure 10 .    Schematics of the confinement of an optical guided for wide, narrow, separate-confinement waveguides.

The medium gain $g_{th}$ at threshold is obtained by stating that the optical wave intensity after a roundtrip in the cavity must stay equal, under the opposite actions of losses and gain. This is conveniently written as :

$$I_o\, R^2 e^{2\left(\Gamma g_{th}\, -\, \alpha\right)L}\, = I_o \qquad (16)$$

where R is the facet reflectivity, $g_{th}$ represents the modal gain per unit length of the optical wave, $\alpha$ sums all the various loss mechanisms (such as free carrier material absorption, waveguide losses...) and L is the laser length. Equation (16) can be rewritten in the form :

$$\Gamma g_{th} = \alpha + 1/L \, \text{Log} \, 1/R \qquad (17)$$

In GaAs/GaAlAs lasers, the first term is usually $10$ cm$^{-1}$ and the second $40$ cm$^{-1}$ for a $300\mu$m long laser and uncoated facet reflectivity of 0.3.

The threshold current is known once the relation between gain and injected current is determined. This can be readily done by using the carrier density as an input parameter. One can both calculate for that carrier density the maximum gain and the required injection current from the known radiative and non-radiative recombination channels. As mentioned above, gain occurs only once a significant inversion has been achieved.

The maximum gain value can be obtained from the calculation of the $\chi''(\omega)$ along equation (5) and (6). Under occupation of states, one has to include in $\chi(\omega)$ all possible transitions, counting as positive those originating in the valence band (upwards, absorptive) and negative those originating in the conduction band (downwards, emissive). By incorporation of the distribution factors $f_c(\omega)$ and $f_v(\omega)$ in conduction bands and valence bands respectively, one finds

$$\chi(\omega) = \frac{1}{\varepsilon_o \hbar} \int_{E_g}^{\alpha} |\langle \psi_c | \vec{er} | \psi_v \rangle|^2 \, \rho_j(\omega_o) \, [f_v(\omega_o) - f_c(\omega_o)] \times \frac{\omega_o - \omega - i\gamma/2}{(\omega_o - \omega)^2 + \gamma^2/4} \, d\omega_o \quad (18)$$

where $f_v$ and $f_c$ are those corresponding to the transition energy $\omega_o$ and $\rho_j(\omega_o)$ is the joint density of states for transitions between $\omega_o$ and $\omega_o + d\omega_o$. After $\omega_o$-integration, one finds the 3D gain as

$$g(\omega) = -\frac{k\chi''(\omega)}{\varepsilon_b} = \frac{|\langle \psi_c | \vec{er} | \psi_v \rangle|^2}{\lambda_o \varepsilon_b \, n\hbar} \left( \frac{2m_r}{\hbar} \right)^{3/2} [f_c(\omega) - f_v(\omega)] \left( \omega - \frac{E_g}{\hbar} \right)^{1/2} \quad (19)$$

which is also

$$g(\omega) = \alpha(\omega) \, [f_c(\omega) - f_v(\omega)] \quad (20)$$

where $\alpha(\omega)$ is the absorption coefficient when no or little carrier injection is present. The condition for net gain, $f_c(\omega) - f_v(\omega) > 1$, yields the usual Bernard-Durrafourg condition

$$E_{Fc} - E_{Fv} > \hbar\omega \quad (21)$$

where $E_{Fc}$ and $E_{Fv}$ arequasi-Fermi levels in conduction and valence bands under injection.

In 2D, the gain formula uses a 2D joint DOS, yielding for a single QW level

$$g(\omega) = \frac{|\langle \psi_c | \vec{er} | \psi_v \rangle|^2 \, 2\pi m_r}{\lambda_o \varepsilon_b \, \hbar^2 L_z} [f_c(\omega) - f_v(\omega)] \quad (22)$$

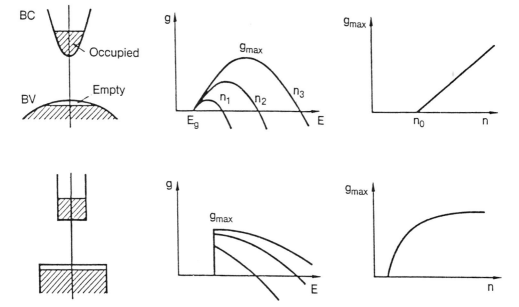

Figure 11 . Schematics of the gain formation in DH lasers (top) and Qw lasers (bottom)

Figure 11 evidences the buildup of spectral gain in 2 and 3D with increasing carrier densities. The remarkable simplifying feature is that in 3D, the gain-current relation, i.e. the variation of the volume gain curve maximum versus injection current is a straight line to an excellent approximation, which can be written as

$$g_{max} = g = A \ (J - J_o) \tag{23}$$

A is therefore the differential gain and $J_o$ the so-called transparency current. $J_o$ represents the injected current needed to reach the carrier inversion : up to $J_o$, carriers do not produce any gain at any energy. Above $J_o$, carriers start to be effective for gain and laser action. It can be inferred that $J_o$ will decrease with active layer thickness.

From equations (17) and (23), one can deduce the current threshold, which becomes, when including an internal quantum efficiency $\eta$ representing the fraction of carrier recombination which is radiative and therefore participates in gain

$$J_{th} = \frac{J_o}{\eta} + \frac{\alpha_i}{\eta A} + \frac{(1 - \Gamma)\alpha_c}{\Gamma \eta A} + \left(\frac{1}{\eta \Gamma A}\right)\left(\frac{1}{2 L}\right) \text{Log}\left(\frac{1}{R_1 R_2}\right) \tag{24}$$

For good GaAs/GaAlAs material, $\eta$ can often be taken as unity, whereas this is far from true for many other materials pairs. The variation of threshold current with DH active layer thickness[34] (figure 12) can then be easily analyzed : Above 1000Å, $J_o$ increases with increasing active layer thickness, like the number of quantum states to be inverted, A decreases because the same current density populates less states per unit volume and $\Gamma$ increases only slightly. As a result, $J_{th}$ increases with d in that region. Below 1000Å, the last term in (24) increases dramatically with decreasing active layer thickness as $\Gamma$ diminishes very fast (as $d^2$) in a DH whereas $J_o$ decreases only as d and A increases only as $d^{-1}$.

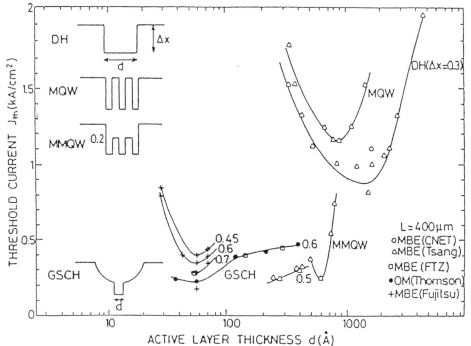

Figure 12 . Overview of current thresholds for various structures (from Noblanc[466])

As is evident from figure 12 , many quantum well lasers can perform much better than DH's. They also display a number of additional advantages which are explained from a more detailed analysis which we outline below for the various structures used (figure 13) which all have their advantages and inconveniences.

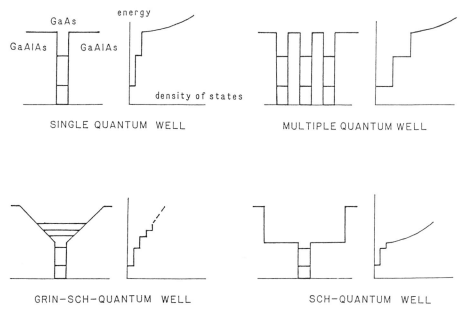

Figure 13 . The various QW structures used as active layers in lasers and the associated density of states

b) Single-quantum well (SQW) laser operation

The use of a very thin active layer has several consequences :

(i) Energy levels in the conduction and valence band become quantized. Therefore, lasing will occur at energies determined by the band gap and the confining energies. Several other effects contribute to the precise energy level position, such as band gap renormalization and Coulomb effects, which will not be discussed here any further[24].

(ii) Due to wavefunction and energy quantizations, the density-of-states becomes two-dimensional. The build-up of gain for three different injected carrier densities in a quantum well is schematically depicted in figure 11 . As can be seen, carriers are more "efficient" than in 3D as added carriers will contribute to gain at its peak (the bottom of the 2D-band), whereas in 3D added carriers move the peak gain away from the bottom of the band, making all carriers at energies below that of $g_{max}$ useless. Therefore, the spectral gain curve in 2D has a larger slope A than in 3D. However, the price to pay is that gain will saturate at a given finite value when the electron and hole states are fully inverted, whereas $g_{max}$ never saturates in 3D due to the filling of an ever increasing density of states. Finally, the transparency density or current is much smaller in a QW than that for a DH laser as the density of states to be inverted is quite smaller (a GaAs quantum well has $\sim 10^{12}$ states per $cm^2$ in the minimal energy range of $kT$ to be inverted instead of $\sim 10^{13}$ states in a 1000 Å DH laser).

(iii) The confinement factor $\Gamma$ has to be kept to an optimized value by using a separate optical confinement scheme of the SCH or GRIN-SCH type. It can be calculated that optimized GRIN-SCH structures yield very comparable values for $\Gamma$ as straight SCH's.

The detailed calculation yield the results shown on figure 14. It also determines an optimal quantum well thickness : at first order, similar injected numbers of carriers above the transparency density have roughly the same efficiency to amplify an optical wave. [This latter point is also true in 1D and 0D : as the optical matrix element is the same for any LDS, (neglecting correlation effects) the transition rate **per allowed electron-hole pair** (i.e. $f_e = 1$, $f_v = 0$) is the same, yielding the same total gain. **What changes with dimensionality is transparency density and the spectral repartition of gain**]. However, the curves do not scale exactly owing to second-order effects : population of higher-lying levels plays a role, in particular for wide wells. Also, for very thin wells, the quasi-Fermi level is so high in the well (due to a large confining energy) that some carriers spill over into the optical confinement cavity, yielding a large $J_0$. These effects explain why one observes an optimum in the layer thickness for minimizing the current threshold. The occupancy of optical cavity states plays a major role in the difference between SCH and GRIN-SCH structures : even in optimized structures, the quasi-Fermi level is so high in the conduction band (in the valence band it is quite lower due to the larger density of states, see below ) that population of the cavity states occurs : at threshold, it can be calculated that the number of electrons in the SCH optical cavity is roughly equal to that of the electrons in the active layer, whereas this number is only about 20 % in the GRIN-SCH structure.

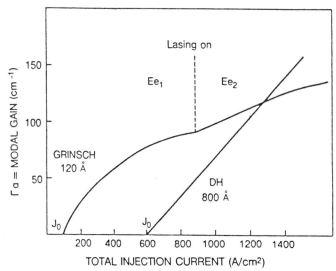

Figure 14 . Gain-current curve for a 120 Å QW GRIN-SCH laser and a 800 Å DH laser, evidencing the changes in transparency currents $J_0$, and differential fains (slopes of the gain-current curves). Note that additional gain is obtained from n = 2 Quantum level inversion at high currents[465].

c) Multiple Quantum Well Lasers

Their properties can easily be deduced from those of a single quantum well if one assumes homogeneous injection of carriers in the various wells : in the approximation where $\Gamma$ can be considered exactly proportional to the total active layer width, the modal gain $\Gamma g$ -current relation can be deduced from that of a single quantum well by simple scaling transformation, i.e. by multiplying units on both graph-axis by the number of wells (figure 15). The advantages and disadvantages of MQW structures are obvious, : whether the SQW or MQW is the better one depends on the loss level : at low loss, the SQW laser is always better because of its lower $J_0$ (only the states of one QW have to be inverted) and lower internal losses ($\Gamma\alpha_i$ scales with the number of wells). At high loss, the MQW is always better because the gain stems from a high-slope part of the gain-current curve instead of the saturated part of the SQW gain-curve. In some cases, the saturated gain of SQW is not large enough and one would absolutely require several QWs to reach the threshold gain. An evident advantage of MQW's is the higher-differential gain which leads, as discussed below, to higher modulation frequencies and narrower linewidth.

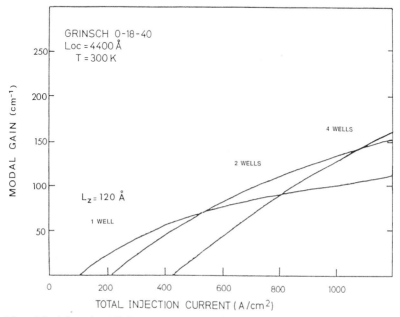

Figure 15 . Modal gain (Gg)-current curve for 1,2,4 well GaAs/Ga$_{0.82}$Al$_{0.18}$As/ Ga$_{0.6}$ Al$_{0.4}$As GRIN-SCH's[462]

The way to ultra-low threshold lasers is also clear[36] : any diminution of the required threshold gain by using reflection coatings on facets (equation 17) will have a major effect in SQW lasers: it will be translated into lower threshold current, which can be made very near its limit $J_0$. The effect of such coatings would be much smaller in DH or MQW lasers, as the major part of the current is the useless $J_0$ in such lasers.

Another path to improvement of the threshold current is the modification of the valence band structure : under usual pumping conditions, the inversion is much smaller in the valence band than in the conduction band as shown in figure 16 (i.e. $f_v(E) \ll f_c(E)$, translating into the fact that $E_{fv}$ is usually quite above the confined valence levels) : this arises from the many nearby valence levels (confinement energy for holes are small due to the rather heavy hole mass) with high density of states, which lead to smaller occupancy factors $f_v(E)$ in order to account for the neutrality conditions

$$N = \int \rho_c(E)\, f_c(E)\, dE = \sum_i \int \rho_{vi}(E)\, f_h(E)\, dE \qquad (25)$$

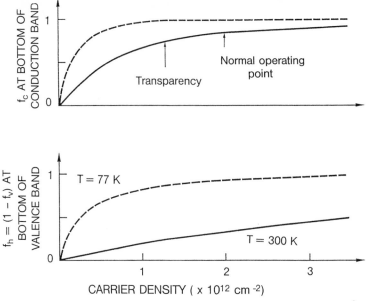

Figure 16. Calculated occupancy-factors at the bottom of bands of a 60Å GRIN-SCH unstrained GaAs-GaAlAs laser as a function of carrier densities, at 300 and 77K. Transparency and normal-operation points are indicated. Note the low occupancy of hole states under 300K operation, due to the numerous nearby hole states

where the summation is carried over all valence bands labelled i, considering for the sake of simplicity that only the ground electron level is populated here.

Being able to split apart the various hole states will significantly increase the inversion in the lowest level at a given hole density, which will lead to higher gain according to equation (19). This effect has been dramatically evidenced by the excellent results obtained on strained-layer pseudomorphic GaInAs/GaAlAs/ GaAs[44] and GaInAsP/InP[45] lasers where the strain splits the valence band, raising a valence band with a light transverse hole mass quite above a heavy-hole mass band. Additional useful features of these heterostructures due to the higher refractive index and deeper electron (hole) quantum wells are better optical confinement factor, higher differential gain (and associated superior lasing characteristics), higher catastrophic damage threshold, transparent substrate for surface emission.

d) The temperature dependence of the laser threshold current

The temperature dependence of current threshold is usually approximated by

$$J_{th}(T_2) = J_{th}(T_1) \exp((T_2 - T_1)/T_0) \tag{26}$$

where $T_1$ and $T_2$ are two nearby temperatures. The variation of threshold current is slightest (which is highly desired in most applications) when the temperature coefficient $T_0$ is larger.

Several effects come into play in the calculation of $T_0$, such as the broadening of the

Fermi-Dirac distribution function, the thermal spillover of carriers above confining barriers etc... A main effect in quantum wells was shown to be the spillover of carriers in the confining layers[40]. This explains why the structures have better (i.e. larger) $T_o$'s in the order $T_o$(SCH) < $T_o$ (GRIN-SCH) < $T_o$ (MQW). The latter is the best because it has the 3D DOS which is the farthest away in energy. It also has a better $T_o$ than the standard DH laser because the high-energy tail of the Fermi-Dirac distribution function only populates a constant 2D DOS instead of populating an energy-increasing 3D DOS with useless carriers. Of course, deeper QW structures such as GaInAs/GaAlAs/GaAs strained QW lasers have improved $T_o$ characteristics, due to both the better confining potential barriers and the lower band filling originating in the more symmetrical electron and hole states described above.

e) Modulation speed and spectral width

Both these parameters, of paramount importance for high-speed telecommunications, are largely improved by the use of QW's as active layers They both depend on the differential gain A = dg/dI which is larger in QW lasers due to the square DOS as discussed above. It can be shown that the relaxation frequency $\Omega_R$ is given by[17]

$$\Omega_R = \left( A\, P_o\, /\, \tau_\rho \right)^{1/2} \tag{27}$$

where $P_o$ is the average laser power and $\tau_p$ the photon lifetime in the laser cavity. As the diminishing of $\tau_p$ increases the cavity losses, the best lasers are MQW's where A is large at large gain, as discussed previously. An additional improvement has been obtained for A by using a p-doped active layer[37]. In that case, the differential gain is larger than for undoped active layers as the quasi-Fermi level for electrons $E_{fc}$ changes very rapidly with injected electron density and inversion is reached as soon as $E_{fe}$ reaches the confined electron level (with the Bernard-Durrafourg relation (21) obeyed) (the undoped or n-type doped cases are less favorable because of the large hole DOS to be inverted). Relaxation oscillations as high as 30 GHz have been reported[37]. Higher frequencies could be reached by reducing the damping factor originating in non-linear gain saturation due to hole burning[38]. Spectral width improvement of QW lasers is by now well-documented[39].

f) Power lasers

The advent of laser-diode pumped solid state lasers opens the way to many applications such as high-power pulses, ultra-narrow linewidth, visible-doubled lasers with high efficiency, compact tunable parametric oscillators etc[40]... At the origin of this explosion is the availability of high-power arrays of laser diodes, which are all based on quantum well structures. The success of QW lasers as generators of high-power is mainly due to two effects[41]:
- Quantum efficiency is very high : the transparency current, corresponding to unused excitation, is smaller for QW lasers.
- Internal losses are smaller in QW lasers because of the small confinement factor.

Therefore the electrical-to-optical power efficiency is better, which is essential as laser arrays are limited by the heat which has to be extracted from the laser die. An additional useful feature is the better catastrophic damage threshold.

g) Vertical lasers

Vertical-emitting lasers have long been desired because of the various applications they would allow : free-space direct chip-to-chip communication, random placement on a chip, easy access for optical accessories (lenses, fibers ...) possibility of 2D-arrays for high powers with low divergence or for 2D imaging. Various schemes have been designed. The most spectacular realization is that with integral mirrors, which has allowed the simultaneous fabrication of more than 1 million lasers on a single chip, with fabrication yields in the 95-100 % range. Current thresholds in the mA have been obtained. Three main factors determine the performance.

(i) mirror reflectivity : due to the limited total thickness allowed by the growth process, the mirror loss factor in a round trip (last term in equation 17) becomes prohibitive for simple semiconductor-air interfaces, and reflective coatings are required. Both dielectric and semiconductor multilayer reflectors have been used (figure 17). The excellent quality of semiconductor multilayer growth has allowed single-step laser fabrication with quarter-wave stack Bragg-reflectors having reflectivities in the 99.96 % range. When compared to the usual 30 % cleaved-facet reflectivity, this allows a $\approx$ 100-fold decrease in cavity length to reach similar photon-losses, which is the usual range of cavity length in vertical lasers ($\sim$ 1-2 $\mu$m). Therefore, such lasers operate with losses (and gain) similar to those of edge-emitting lasers.

Figure 17 .   Micrograph of an 2$\mu$m-wide etched mesa vertical laser with integral Bragg GaAs-GaAlAs multilayer reflector. Structure has been overetched to show the alternate GaAs/GaAlAs plans[42] (courtesy J.P. Harbison, Bell Core)

(ii) Lateral confinement is required in order to reduce the active volume. For wide lasers, only current confinement is required, whereas below the $\mu$m diameter range, optical confinement is also required, like in standard edge-emitting DH lasers. Several schemes have been used : substrate etching defines a tight optical beam, while current confinement is obtained through lateral p-n junction barrier confinement. Conversely, etching of the mesas through the mirrors and active layers also defines excellent tight optical beam and confined-current injection. However, this is at the expense of large surface leakage currents created by the mesa fabrication through reactive ion-etching. Therefore, instead of a decreasing threshold current with mesa-diameter (at least down to 0.1μm, from which G would start to decrease too much), one observes an optimal-threshold current for 5 μm-diameter mesas, below which the diffusion current to the damaged surfaces takes over the standard radiative recombination currents.

(iii) Carrier injection usually occurs through the mirror materials, which have a large series resistance, in the range of several tens of ohms. This leads to severe laser heating under c.w. operation, and is a major limitation for designs where current flows through the mirrors.

The results achieved so far are however impressive and are most often obtained on GaInAs strained-layer QWL's[43] : sub-milliamp threshold, 8 GHz operation, 30 ps pulsed-output. The losses are in the 10 cm$^{-1}$ range which shows that dominant losses are cavity losses. The electrical-to-optical power efficiencies are in the 10 % range. Power output is limited both by the maximum intensity acceptable on mirrors ($\sim 10^7$ W.cm$^{-2}$) and required device cooling.


## IV - ELECTRO-AND NON-LINEAR -OPTICAL PROPERTIES AND APPLICATIONS

a) Basic principles[17,44,45]

Electro-optical and non-linear -optical phenomena are well described by explanding the electric polarization P in powers of applied electric field E as

$$P = \varepsilon_o \left( \chi^1 E + \chi^2 E^2 + \chi^3 E^3 + ... \right) \qquad (28)$$

The index of refraction is then

$$n^2 = (n_r - i \, n_i)^2 = \left( 1 + \chi^1 + \chi^2 E + \chi^3 E^2 + ... \right) \qquad (29)$$

Generally, the $\chi$'s are tensors and $\chi^2$ and $\chi^3$ lead, for an harmonic electric field, to second and third harmonic generation of light. When $\vec{E}$ is the sum of DC and optical fields such as $\vec{E} = \vec{E}_{DC} + \vec{E}_o \exp i (\omega t - kz)$ various effects can happen, as various frequencies will be generated. Using the usual notation, $\chi^n \left( \omega_R = \omega^1 + ... + \omega^n ; \omega_1, \omega_2 ... \omega_n \right)$, one of the $\chi^2$ terms will be, at zero frequency $\chi^2 ( o ; \omega, -\omega) E_o^2$. It represents the optical rectification phenomena. Another will be $\chi^2 (\omega ; o, \omega) E_{DC} E_0$, the usual linear Pockels effect, which describes the linear action of a DC electric field on a non-centrosymetric material. The term $\chi^3$ $(\omega ; o, o, \omega) E^2_{DC} E_0$ describes the DC Kerr effect, or quadratic electro-optic effect, which exists in all materials, but is usually negligible whenever a linear effect is present.

Classical[17] and quantum calculations[44,45] of the various $\chi$'s can be done. In the latter case, perturbation theory will quite easily yield the required expressions. They involve a number of terms which quickly become cumbersone, and one usually only retains those which are near-resonant. It is however within easy grasp that $\chi^2$ expressions will contain terms of the type

$$\frac{\langle 1 | \vec{er} | 2 \rangle \langle 2 | \vec{er} | 3 \rangle \langle 3 | \vec{er} | 1 \rangle}{\left( (\omega - \omega_{12})^2 + \gamma_{12}^2 / 4 \right) \left( (\omega - \omega_{13})^2 + \gamma_{13}^2 / 4 \right)} \qquad (30)$$

where |1>, |2>, |3> label different quantum states, while $\chi^3$ will contain terms with products of four matrix elements and three resonant energy denominators.

What we require from this analysis is the simple remark that whenever $\chi^1$ will be large in LDS's due to various reasons described above for QW's and below for lower-dimensionality systems, $\chi^2$'s and $\chi^3$'s will also be large as they incorporate similar oscillator strengths and resonances. This is usually known in the field of non-linear optics as Miller's empirical rule[46]. By defining quantities such as

$$\Delta = \frac{\chi^2(\omega_1 + \omega_2)}{\chi^1(\omega_1 + \omega_2)\ \chi^1(\omega_1)\ \chi^1(\omega_2)} \tag{31}$$

Miller found that D was approximately constant for a wide family of solids in which $\chi^2$ varies by four orders of magnitude.

Several non-linear resonant calculations have indeed been performed for quantum wells[24,47]. However, whenever resonant excitations are used, leading to finite state population, quantum perturbation theory cannot be applied. Exact calculations are possible for isolated atoms where one develops saturation models, mainly in the resonant two-level approximation. In solids, completely correct descriptions are more difficult as evolution of state populations is complicated due to the numerous intensity dependent- processes (elastic and inelastic collisions, lifetime etc...) (figure 18). One therefore resorts to simpler phenomenological models, where the modification of quantum states and their occupancy under applied excitation is first calculated or measured, and one then calculates or measures the ensuying modification of optical constants. Expressions like (30) are however very useful to determine scaling laws and to compare different systems.

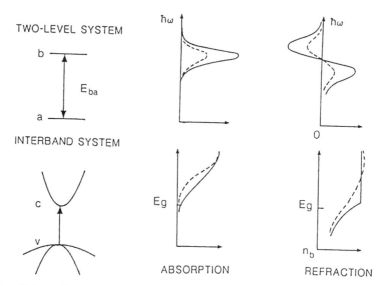

Figure 18 . Schematics of the energy levels and changes in absorption and refraction index under strong optical real excitation for a discrete two-level system (top) and continuum interband system (bottom). From Wherett[47]

The electro-optic effects are therefore calculated in two-steps : first the effect of an applied electric field on the bandedge absorption spectrum through QCSE or through PSF is calculated or measured. From the induced change in the absorption coefficient $\Delta\alpha$, i.e. $\Delta n_i$ or $\Delta\chi''$, one deduces the other required quantities such as the changes $\Delta n_r$ or $\Delta\chi'$.

In a similar manner the changes due to intense light beams ($\chi^3$ term, Kerr effect) are usually calculated by first evaluating the induced-changes in states populations, and then deducing the resulting $\Delta\alpha$, $\Delta v$, $\Delta\chi$[1]. Useful parameters are the non-linear index of refraction $n_2$ and absorption coefficient $\alpha_2$, described by

$$n = n_o + n_2\,I \qquad \alpha = \alpha_o + \alpha_2\,I \qquad (32)$$

where I is the light intensity, and the change per carrier of n and $\alpha$ defined by

$$n = n_o - \eta N\,. \quad \alpha = \alpha_o - \sigma N \qquad (33)$$

where N is the carrier density. Simple relationship exist between $\eta$, $n_2$ and $\alpha_2$, $\sigma$, which include the lifetime $\tau$.

As can be inferred, the outstanding properties of quantum wells stem from several effects which appear separately or together, depending on the particular physical system :
(i) the increased light-matter interaction, due to exciton effects
(ii) the wavefunction confinement, such as in the QCSE
(iii) the diminished accessible phase-space, such as in PSF effects.

b) Electro-optic effects[48]

The two electro-optic effects mainly used, QCSE or PSF, were displayed on figures 8 and 9 (right). Implementation can be used in two geometries, with light beam propagating perpendicular to the semiconductor surface or parallel to it (usual integrated optics geometry). The large efficiency of QW's allows for required interaction lengths (defined by $\Delta n\,\Delta l \sim 1$ or $\Delta\alpha\Delta l \sim 1$) much smaller than for other materials, making the perpendicular geometry possible or the size of integrated optics very compact. In the former case this allows the use of QW's as efficient 2D spatial light modulators ; in the latter case one can fabricate or integrate much more devices on a single wafer, drastically diminishing single-device cost.

The perpendicular geometry has first been used with simple p-i-n transmission structures evidencing excellent response times in the subnanosecond range, limited only by the RC-time constant of the electrical circuit[48]. This is normal as in the operating mode where no or few carriers are created like in the QCSE, the intrinsic response time of the material is the very fast dielectric response time which describes the propagation of the applied electric field in the structure, the transient electronic and ionic motions being even faster. The energy requirements for switching are also excellent : For a typical 10µm x 10µm x 1µm (thickness) modulator, the capacitance is $\approx$ 12fF, which yields a commutation energy $\approx (1/2)CV^2 \approx 150$ fJ at 5V drive. One both observes that this is not quite as good as modern purely electronic devices (for a 1 µm technology MOSFET, P. $\tau \sim 10$ fJ), but at the same time this should be compared to an input/output device of an electronic chip, as the electro-optic modulator is directly connected to the outside world. The comparison is then much better for the QW modulator, as it is well-known that interconnection wires and pads on chips are much larger than individual logic cells and therefore require high driving currents (therefore large switching energies). This direct connection to the outside world is one of the advantages of optical interconnects and computing, besides the high-degree of parallelism allowed by the 3D geometry of optical systems.

The main drawback of the early devices was their large insertion loss and poor contrast, due to their operation in the high-absorption region of QW's. This situation has however changed dramatically in the recent past due to much better implementations of QW's for electro-optic applications. One should remember that multilayered-dielectric mirrors can be tailored to yield excellent reflection or transmission factors by a proper choice of phase factors like in Bragg or Fabry-Perot structures, although individual starting materials have poor performance. Both Bragg reflectors[49] and Asymetric Fabry-Perot (ASFP)[50] MQW modulators have been designed and demonstrated, with superior performance in the reflecting mode (contrast ratio > 100). This is obtained in the ASFP device because the reflection coefficients of the two multilayered mirrors and phase can be tailored so that the back-mirror reflected beam exactly cancels the front-mirror reflected beam, leading to excellent extinction. The Bragg reflectors devices mainly use the large changes in absorption coefficient to strongly modify the propagation / reflection of light, whereas the ASFP uses both changes in phase (index) and intensity (absorption) to modify light beams. A recent ASFP implementation uses a superlattice as the active medium and its absorption change due to Wannier-Stark localization under an applied electric field[51].

In the parallel, integrated optics geometry progress is also important. In a series of experiments, Zucker et al. have measured and compared QCSE[52] and PSF[53] - based devices in the ternary / quaternary GaInAsP materials case, both in intensity and phase-modulated devices. The change in index of refraction in the QCSE is usually given as $\Delta n = - (1/2) n^3 [ rE + s E^2]$, where r retains the standard 3D value of the linear electro-optic coefficient. The QCSE value of s ($-0.15 \times 10^{-14}$ cm$^2$ V$^{-2}$) is typically 100 times larger than that due to the Franz-Keldysh effect in the bulk and allows sub-mm sized interferometers. PSF devices even evidence better properties : the waveguide effective index change per volt is $\Delta n / V = 2.2 \times 10^{-4}$ V$^{-1}$ at 1.58 $\mu$m, allowing 180° phase shift voltage of 5.4 V in a 650 $\mu$m long Mach-Zehnder interferometer. In addition, the PSF devices have a large $\Delta n / \Delta \alpha$ ratio ($\Delta \alpha$ : electro-absorption), which is desirable for those devices which require a pure phase modulation without intensity modulation.

## c) Non-linear effects

Non-linear are usually measured through pump and probe experiments (fig. 9 left), yielding $\Delta \alpha$ from which $\Delta n$ is calculated by the Kramers-Kroning integral. The excellent non-linear properties of quantum wells are evidenced by degenerate four-wave mixing with reasonable diffraction efficiencies at extremely low pump intensities, such as obtained from a semiconductor laser diode[54] : $5.10^{-5}$ for a 1.25 $\mu$m thick sample at light intensity of 17 W cm$^{-2}$. This clearly opens the way to full optical signal processing with quantum wells as materials allowing light-by-light beam steering.

Room temperature absorption saturation experiments also evidence superior performance of quantum wells. The saturation intensity $I_s$ is defined from a variant of equation (33) as

$$\alpha = \alpha_0 \left(1 + \frac{I}{Is}\right) - 1 \qquad (34)$$

In all non-linear measurements, one finds a saturation intensity defined by a critical (saturation) carrier density $n_{sat} \sim \left( \pi\, a_B^2 \right)^{-1}$, the areal density of closely packed excitons. This is quite normal as, whatever mechanism is chosen (real space or phase-space filling), this quantity is also (with $L \sim a_B$ in usual quantum wells)

$$n_{sat} \approx \rho_{2D} \times kT \left( \approx \left( \frac{m}{\pi\hbar^2} \right) \left( \frac{\pi^2\hbar^2}{2\,mL^2} \right) \right) \qquad (35)$$

corresponding to phase-space filling of all states up to the energy $kT$, including those situated at the Rydberg energy from which the exciton states are constructed.The dependence of saturation intensity on well width has been directly traced to the exciton diameter, evidencing a maximum when that diameter is minimal, i.e. at $L \sim 75\text{Å}$[55].

When dealing with **intersubband transitions** in the infrared, QW properties are significantly enhanced. As was described in figure 2, QW's act as artificially-large molecules with giant electric dipoles. This leads to infrared unsurpassed non-linear properties. The calculations are enormously simplified for the intersubband transitions : A simple two-level atom model can be used in the density-matrix formation of non-linear coefficient by injecting their carrier density N (due to intentional doping) as the number of atoms in the linear and non-linear susceptibility formulas. This simplification is due to the fact that the band dispersion (and therefore the DOS) does not play a role as transitions occur between particles (electrons or holes) with the same dispersion, i.e. transitions occur at the same energy independently of the electron wavevector. The optical rectification coefficient at resonance can then be shown to be[56]

$$\chi^2 (0) = 2e^3 \left( T_1\, T_2\, /\, \varepsilon_o\, h^2 \right)\, N\, |\langle 1|z|2 \rangle|^2 [\, \langle 2|z|\, 2 \rangle - \langle 1|z|1 \rangle\, ]\, \sin^2 \theta \qquad (36)$$

where $T_1$ is the longitudinal relaxation time, (usually the state lifetime), $T_2$ the transverse relaxation time ($T_2 = \gamma^{-1}$), the bracket term represents the dipolar charge-build up between excited and ground state, $\sin\theta$ is the intersubband selection rule factor involving the internal light beam angle with the normal to the layers. Simple asymetric quantum wells (figure 19a) yield values for $\chi^2(0)$ about $10^3$ larger than in bulk GaAs[57]. More recently, the use of specially designed coupled quantum wells has given a further improvement of 300, by increasing the lifetime $T_1$ of state 2 (usually a subpicosecond L0-phonon time in the structure of figure 19a), using a spatially-separated metastable state which can only tunnel back to the ground state, and by also increasing the dipolar charge build-up by wide well separation[58] (figure 19b).

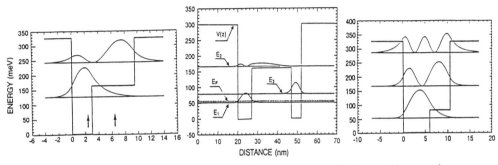

Figure 19 .   Various composition-modulated quantum wells for non-linear optics

Second-harmonic generation can also be extremely efficient, due to the large oscillator strengths and the double-resonance situation which can be obtained through careful energy-level design. In that case, one calculate a second harmonic generation coefficient $\chi^2(2\omega)$ given by[56,59]

$$\chi^2(2\omega) = \frac{e^3 N}{\varepsilon_0 h^2} \frac{\langle 1|z|2\rangle \langle 2|z|3\rangle \langle 3|z|1\rangle}{(\omega - \omega_{12} - i\gamma_2)(2\omega - 2\omega_{13} - i\gamma_3)} \qquad (37)$$

Such theory and experiment yield at resonance a value $\approx 10^3$ larger than in bulk GaAs for the structure shown in figure 19c. From formula (37) it appears that an asymetry is required to obtain non-zero $\chi^2$ through dipole selection rules. Such an asymetry is obtained through compositional asymetry like in figure 19c or through an applied electric field on a symetric quantum well structure.

The extraordinary efficiency of QW's is well evidenced by the fact that non-linear generation can be obtained with a low-power $CO_2$ cw laser (300 W cm-2) on a 1μm active length (the sample thickness) !

So far, the non-linear properties of QW's based on interband transitions have remained a subject for physics studies, as the required power levels are still rather high, although better than those of other non-linear materials. However, by using both the photodetector properties of QW's along with their electro-optic modulation properties, one can design an efficient low-power non-linear optical device, thanks to external electrical circuitry which provides an efficient amplifying feedback. Usually, such systems are made in a hybrid form . Here, MQW's can perform both photodetector and modulator functions. A resistor Self-Electro-Optic-Effect Device (SEED) is shown on figure 20 as used as a bistable device[75]. The MQW's are imbedded in a p-n junction and serve both as photodetectors and modulators. The incoming photon energy is at the dashed-line position in the absorption spectrum. When no light is shining, all the voltage (10 V) is applied to the structure, resulting in a small absorption. As light starts to increase, a voltage drop occurs across the resistor due to the photocurrent, reducing the voltage applied to the MQW's therefore increasing absorption. Passed some critical

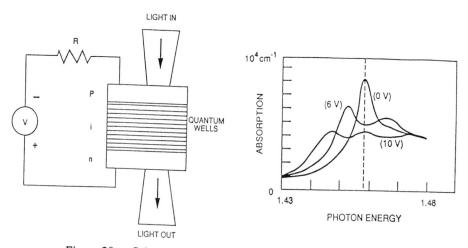

Figure 20 . Schematics of SEED biasing and operation (from 48)

"switch-down" energy, this process will lead to a runaway which puts the diode in a high absorption, high-current, low-applied voltage mode. The critical energy which leads to a non-linear absorption behavior is of course much smaller than the one required in a fully optical non-linear effect, as the modification of the absorption edge is provided by an electro-optical effect which only requires a capacitor charging from an external c.w. electrical source, instead of a phase-space filling[76] :

$$P_{in} \to P_{abs} \to N \to I_{ph} \to P_e \to \Delta F \to \Delta E \to \Delta \alpha$$

where $P_{in}$ is the incoming optical power, $P_{abs}$ the absorbed power, N the photocreated current density, $I_{ph}$ the photocurrent, $P_e$ the energy supplied by the electrical generator, $\Delta F$ the change in electric field applied to the MQW's, $\Delta E$ the change in MQW band-edges due to the QCSE, $\Delta \alpha$ the resulting change in absorption coefficient.

The SEED device is bistable as switching to the high-voltage, low-current, low-absorption regime will occur at a lower "switch-up" optical power than at "switch-down". Switching power and speed can be adjusted by the resistor load as the circuit time constant is $\approx$ RC, C being the capacitance of the electro-optic device. The optical switching energy is $\approx 1/2$ $CV^2$, as switching occurs when the photocreated charge $\approx$ cancels the charge stored in the device acting as a capacitance i.e. screens out the applied electric field. The switching energy is therefore $\approx 1.7 \ 10^{-7} \ J.cm^{-2}$ which is $\approx 100$ times smaller than the energy required to produce an equivalent absorption change in multiple quantum wells through the straight non-linear absorption.

Many implementations of SEED devices have been demonstrated[48]. One of the drawbacks of the simple SEED is the lack of intrinsic gain which require the device to be operated near the switching threshold so that some amplification is obtained as required to be able to cascade devices. This near-threshold operation also makes the device very sensitive to any fabrication fluctuation, spurious light illumination etc... To overcome this limitation, the symmetric SEED (S-SEED)[60] was developed. One should consult recent reviews of this field to both evaluate the potential of such devices and their already amazing state of performance integration of 2000 devices, 40 pW holding power per device, switching speed below 1ns[48].

## V - BASICS AND APPLICATIONS OF QUANTUM WIRES AND QUANTUM BOXES

Considering the huge success of physical systems and devices based on 2D heterostructures, it is natural to explore lower dimensionality systems with the hope to obtain still newer phenomena ad better devices. We will here outline some of the foreseeable new optical physics and devices leaving the fabrication principles and energy level calculations as they are covered in Professor Beaumont's lectures[61].

If one restrict particles to a narrow line (so-called Quantum Wire Well QWW) or narrow box (Quantum Box QB), further quantization appears as compared to 2D QW's. The reasons to expect new phenomena or properties will stem from the same effects which determined QW properties :

(i) wavefunction confinement and localization

(ii) better k-matching between electron and hole states and smaller density-of-states

(iii) large modifications induced by applied fields or band-filling

We will describe these various effects as they apply to QWW's or QB's

a) Wavefunction confinement and localization

The first parameter which has to be evaluated is the oscillator strength. For intersubband transitions, giant oscillator strengths will occur, similar to those observed in 2D, with the photocreated electric dipole oriented along the exciting electric field polarization. An interesting feature is that the selection rule which requires the light polarization to be along the intersubband dipole moment will now be possible in a perpendicular propagation geometry ( E parallel to surface), whereas the only useful component of the E-field in QW's was the perpendicular one.

Neglecting Coulomb correlation effects, the interband oscillator strength is unchanged in 1D and 0D, like in 2D as integration of the matrix element in the three directions by separation of fast varying and slow varying functions in the integral will yield results similar to those obtained in 2D. Taking now into account the Coulomb interaction between electron and holes, several situations can develop, depending on the relative sizes of the Rydberg energy, the confining energy of excitons, of electrons and holes, of kT[62-64].

(i) For very wide QWW's or QB's, where $L > \left(\pi^2 \hbar / 2m_x kT\right)^{1/2}$, quantized exciton levels are separated by less than kT. In that case, exciton levels are mixed and no confinement effect exists.

(ii) For wide QWW's or QB's where B, $a_B < L < \left(\pi^2 \hbar / 2m_x kT\right)^{1/2}$ only the ground exciton state will be populated at temperature T. In that case, the whole box volume (or wire surface) can be excited coherently as a whole. In that case, a giant oscillator strength develops, where in the box case one has an oscillator strength

$$f_{X\ BOX} = \frac{V_{BOX}}{V_X} f_{at}$$

where $V_{BOX}$ and $V_X$ are the QB and free exciton volumes respectively.

(iii) if $L \lesssim a_B$, confinement energies for electrons and/or holes are larger than the Coulomb interaction, which then only appears as a perturbation to the confining potential. In that case electron and hole wavefunctions are fully determined by confinement, and the oscillator strength of the QB is $f_{at}$. Coulomb interaction only shifts the ground-state energy.

The transition between the various cases is of course progressive. QWW's have properties which will be intermediate between QW's and QB's.

Like for QW's, the increased confinement of QWW's and QB's should lead to an increased e-h Coulomb interaction. The ensuying increase in exciton binding energy has been evidenced through magneto-luminescence measurements[65]. Conditions (ii) are difficult to simultaneoulsy fulfill, as satisfying the condition for exciton confinement one usually yields $a_B \sim L$ (case iii). One requires very small Bohr radii, which is only the case for CuCl ($a_B \sim 7$Å). In that case, the giant oscillator strength has indeed been observed through the shortening of exciton lifetime with increased CuCl nanosphere diameter[66,67] (figure 21), as well as the ensuying non-linear $\chi^2$[68]. In CdS, where one should be in an intermediate case ($a_B \sim 25$Å), excitons are readily dissociated when one approches the exciton quantization condition (i) : In the system studied so far, that of $CdS_xSe_{1-x}$ nanospheres imbedded in glass matrices, trapping of electron-hole pairs at interface traps create electric fields which dissociate excitons when the sphere diameter becomes smaller than 300Å. The luminescence diminishes strongly in that range and switches to a donor-acceptor pair recombination mechanism[69].

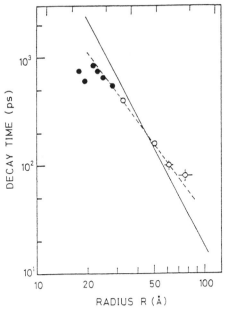

Figure 21 . Variation of photoluminescence lifetime in CuCl nanosphere-doped glasses with diameter, showing the giant oscillator strength increase with sphere size[66]

Many other effects due to new selection rules have been evidenced. As discussed above (see section II) ground-state heavy-hole emission is emitted preferentially with polarization along the longest dimension, while conversely light-hole emission occurs at higher energies appears with a perpendicular polarization. In horizontally-structured materials like QWW's and QB's, important grating-coupling effects appear and great care must be exerted to interpret data. These effects have been studied in a definitive manner by Heitmann and his group[70]. Carrier localization can also manifest itself by more subtle effects. For CdS nanospheres, it appears that Auger recombination is increased, probably due to the break down of strict k-selection rule due to k-spreading[71].

QB's offer also a unique situation in which carriers are localized radiative states. The fact that usual transport does not occur could prove very useful in those instances where carriers usually encounter non-radiative defects. It might well be that the perfect GaAs-on-Si laser will be a QB laser. It is a very simple calculation that the probability of encounter through a diffusive random walk with a dislocation diminishes drastically with dimensionality, reaching almost zero for QB's unless unphysically large dislocation densities are assumed.

b) Smaller density of states

As was discussed in section II, optical properties are determined by $\chi$'s. Whether a smaller density-of-states is useful depends on the fact that sizeable optical effects still exist. The diminution in states allowed by using QWW's or QB's compared to QW's (with its 2D DOS of $\sim 10^{13}$ states cm$^{-2}$eV$^{-1}$) is only useful if these states have larger oscillator strengths and/or more monoenergetic distributions. This latter point will be discussed below as it is linked also to the increasing k-matching of lower dimensionality systems. The oscillator strength of QWW's and QB's should in most cases be compared to that of excitons within QW's, at least for those properties which involve inrelaxed crystal excitations such as those of electro-and non-linear optical devices. In that case, one sees that the unit surface oscillator strength of excitons being $\approx f_{at} \, 8 \, / \, \pi a_B^2$, one needs quantum boxes with mean distance shorter than $\approx a_B$ to enhance the areal oscillator strength, unless these QB's have some emission coherence mechanism.

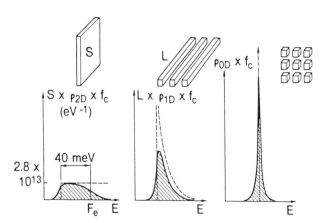

Figure 22 .   Schematics of gain curves in 2D, 1D, 0D structures. Similar numbers of electrons and holes are being injected above transparency, yielding equal integrated gain[465]

c) Matching of electron and hole states

This effect is readily evidenced on the reduced density-of-states, which in turn also represents the spectral repartitions of oscillator strength (figure 3). The ultimate system, the QB's, therefore yields in the solid state "quasi-atoms" which should display atom-like resonances in their optical properties, the density of such atoms being that of the QB's.

The consequences of k-matching are well-examplified in the case of the laser application. Remembering that a k-matched e-h pair yields the same gain in 3, 2, 1, 0D as it has the same oscillator strength (neglecting Coulomb correlation effects) injecting a similar number of e-h pairs above transparency will yield different gains in 2, 1, 0D as shown in figure 22. As can be seen, depending on energy level broadening, e-h pairs are significantly more efficient in 1 and 0D than in 2D. Assuming for instance a linewidth of $\approx 5$meV, one obtains a 8-fold increase in gain for QB's for a given number of carriers. This is somewhat compounded by the symetrization of conduction and valence bands due to stronger quantization in the valence band. If the optical cavity losses have the usual value ($\approx 40 \text{ cm}^{-1}$), one then requires $\approx 10^{11}$ QB's x cm$^{-2}$ since $\approx 10^{12}$ e-h pairs x cm$^{-2}$ were required in 2D (as deduced from laser calculation and measurements).

196

At this stage, it is worth mentioning that the quantum dot laser should bring to a close the evolution of the active material of semiconductor lasers : the DH structure optimized 3D behaviour by simultaneous concentration of the optical wave and carriers. The single quantum well SCH laser diminished almost to perfection (under usual loss factors, i.e. non AR-coated facets) the number of states to be inverted, with additional bonus due to the high-differential gain provided by 2D behaviour. Strained-layers led to symetric conduction and valence bands, optimizing the injection current required to satisfy the Bernard-Durafourg condition. The Quantum Box laser finally squeezes the gain in a narrow spectral range, bringing back the semiconductor laser operation to that of atomic vapor-based lasers, with the essential advantage of compactness and direct current injection.

Various detailed simulations have been performed[72-75]. A major optimization parameter is to keep constant the optical confinement factor by using a SCH configuration. Extremely low threshold currents should be obtained (in the 10μA range) by careful optimization of the lateral confinement ($\approx$ in the 1000 Å range instead of the present μm range). So far, the few experiments reported do not yield such extreme performance, due to fabrication limits and difficulties to achieve a good confinement factor[76].

Like in the QW laser case, other properties than threshold current will also be improved. In particular, modulation speed and spectral linewidth will progress due to the quite better differential gain (up to ~ 10 times better in QB's than in QW's)[75].

Some simulations of QWW and QB lasers have been performed through the 1D and 0D Landau quantization of energy levels obtained when respectively placing 3D DH lasers and 2D QW lasers[76] in a strong magnetic field. Although the reduction in linewidth and increased modulation speed were observed[78], no clearcut diminition of threshold current could be evidenced[79], which might be either due to technological problems or to more intrinsic causes such as diminished relaxation efficiency to the recombination levels.

d) Increased action of external causes

The effect of electric fields should not strongly depend on dimensionality below 2, as the Stark effect is mainly felt in the field direction. Increased electro-optical parameter will therefore only originate from other causes, such as sharper resonances due to k-matching, increased oscillator strength when present, or increased exciton binding energy. Simulation points out to the excellent properties which should be obtained at resonance due to the concentration of oscillator strength[80].

Non-linear optical properties have been widely studied, both theoretically and experimentally. One expects increased effects due to confinement, as occupancy of electron boxes by more than one e-h pair is strongly affected by the Coulomb interaction between carriers. This has been alternately described as biexciton effects[81] or phase-space filling[82]. These effects lead to modified spectral features of the two-photon spectra as compared to the linear optical spectrum. Due to the necessary balance between absorption probability and state filling in usual non-linear effects no clearcut improvement in non-linear threshold is expected (in other terms quantized states are as efficient in 2, 1, 0D when excitons are present), except whenever increased oscillator strength occurs. This has recent been demonstrated in the CuCl nanosphere system, where $\chi^2$ is shown to increase with sphere diameter[68].

ACKNOWLEDGMENTS

It is a pleasure to thank E. Rosencher for useful discussions and a critical reading of the manuscript and B. Marchalot for expert help in production of the text.

REFERENCES

1   M.D. Sturge and M.-H. Meynadier, J.Lumin. 44:199 (1989)
2   G. Bastard, these proceedings, p.
3   G. Bastard, "Wave Mechanics Applied to Semiconductor Heterostructures", Editions de Physique, Les Ulis (France) (1987)
4   L. Sham, Y.T. Lu, J. Lumin. 44:207 (1989)
5   T.P. Pearsall, J. Lumin. 44:367 (1989)
6   G. Abstreiter, these proceedings, p.
7   J.M. Berroir, Y. Guldner and M. Voos, IEEE J. Quantum Electron. QE-22:1793 (1986)
8   J.K. Furdyna and J. Kossut eds, "Dilute Magnetic Semiconductors", Semicond. and Semimetals 25, Academic, Boston, 1988
9   T.P. Pearsall ed., "Strained-Layer Superlattices", Semiconductors and Semimetals, Academic, Boston, 1991
10  see, e.g., Topics in Applied Physics, vol.66, M. Cardona and G. Günterhodt eds., "Light Scattering in solids V : Superlattices and other Microstructures". See in particular chapt. 4 : Spectroscopy of Free Carrier Excitations in Semiconductor Quantum Wells, A. Pinczuk and G. Abstreiter, p. 153, Springer, Berlin (1989)
11  B. Hamilton, this volume p.
12  J.L. Merz, this volume p.
13  J. Shah, IEEE J. Quantum Electron. QE-22:1728 (1986)
14  see, e.g. the special issue of IEEE J. Quantum Electron. QE-25: December 1989
15  see, e.g., R. Dalven, "Introduction to Applied Solid State Physics", Plenum, New-York, (1980), 2nd ed., (1990)
16  Y.R. Shen, "The Principles of Nonlinear Optics, Wiley", New-York (1984)
17  A. Yariv, "Quantum Electronics", 3rd ed., Wiley, New-York (1989)
18  see, e.g., F. Stern, Elementary Theory of the Optical Properties of Solids, in Solid State Physics 15, F. Seitz and D. Turnbull eds., Academic, New-York (1963)
19  M. Asada, Y. Miyamoto and Y. Suematsu, IEEE J. Quantum Electronics QE-22:1915 (1986)
20  L.C. West and S.J. Eglash, Appl. Phys. Lett. 46:1156 (1985)
21  B.F. Levine, C.G. Bethea, K.K. Choi, J. Walker and R.J. Malik, J. Appl. Phys. 64, 1591 (1988) and references therein
22  see e.g., R. Knox, Theory of Excitons, Academic, New-York, 1963 ; J.O. Dimmock, in Semiconductors and Semimetals, vol.3, R.K. Willardson and A.C. Beer eds., Academic, New-York (1967)
23  see, e.g., D.S. Chemla and D.A.B. Miller, J. Opt. Soc. Am. B2:1155 (1985)
24  S. Schmitt-Rink, D.S. Chemla and D.A.B. Miller, Adv. Physics 38:89 (1989)
25  Y. Merle d'Aubigné, H. Mariette, N. Magnea, H. Tuffigo, R.T. Cox, G. Lentz, Le Si Dang, J.L. Pautrat and A. Wasiela, J. Cryst. Growth (1990)

26    M. Asada, A. Kameyama and Y. Suematsu, IEEE J. Quantum Electronics, QE-20:745 (1984)

27    C. Weisbuch, R.C. Miller, R. Dingle, A.C. Gossard and W. Wiegmann, Solid State Commun. 37:219 (1981)

28    C. Weisbuch, in "Physics and Applications of Quantum Wells and Superlattices", E. Mendez and K. von Klitzing eds., NATO ASI Series B : vol.170, Plenum, New-York, (1978)

29    B. Hamilton, this volume p.

30    P.M. Petroff, R.C. Miller, A.C. Gossard and W.Wiegmann, Appl. Phys. Lett.44:217 (1984)

31    P.M. Petroff, C. Weisbuch, R. Dingle, A.C. Gossard and W. Wiegmann, Appl. Phys. Lett. 38:965 (1981)

32    C.V. Shank, R.L. Fork, R. Yen, J. Shah, B.I. Greene, A.C. Gossard and C. Weisbuch, Solid State Commun. 47:981 (1983)

33    H. Sakaki and H. Yoshimura, in "Optical Switching in Low-Dimensional Systems", H. Haug and L. Banyai eds., NATO ASI series B : Physics vol.194, Plenum, New-York, 1988, p.25

34    J.P. Noblanc, Surf. Sci. 168:847 (1986)

35    P.L. Derry, A. Yariv, K. Y. Lau, N. Bar-Chaim, K. Lee and J. Rosenberg, Appl. Phys. Lett. 50:1773 (1987)

36    J. Nagle, S.D. Hersee, M. Krakowski, T. Weil and C. Weisbuch, Appl. Phys. Lett. 49: 1325 (1986)

37    K. Uomi, Jpn. J. Appl. Phys. 39, 81 (1990)

38    K. Uomi and N. Chimone, Jpn. J. Appl.Phys. 28, L1424 (1989)

39    S. Takamo, T. Sakaki, H. Yamada, M. Kitamura and I. Mito Electron. Lett. 25:357 (1989)

40    see, e.g., R.L. Byer, Laser Focus World, March 1990.

41    W. Streifer, D.R. Scifres, G.L. Harnagel, D.F. Welch, J. Berger and M. Sakamoto, IEEE J. Quantum Electronics QE-24:883 (1989)

42    J.L. Jewell et al. Electron.Lett. 25, 1124 (1989) 25:1377 (1989) ; Appl.Phys.Lett. 55: 2724 (1989)

43    J.L. Jewell et al., CLEO 90, Tech.Dig.Ser. 7 (OSA, Washington, D.C., 1990) p. 500

44    Y.R. Shen, "The Principles of Non-Linear Optics", Wiley, new-York (1984)

45    P.N. Butcher and D. Cotter, "the Elements of Non-Linear Optics, Cambridge University Press", Cambridge (1990)

46    R.C. Miller, Appl. Phys. Lett. 5, 17 (1964)

47    B.S. Wherett in "Non-Linear Optics : Materials and Devices", C. Flytzannis and J.-L. Oudar eds, Springer, Berlin (1985), p. 180

48    D.A.B. Miller, Optics and Photonic News, Feb. 1990, P.7 ; Optical and Quantum Electronics 22:S61 (1990)

49    R.-H. Yan, R.J. Simes and L.A. Coldren, IEEE J. Quantum Electron. QE-25:2272 (1989)

50    M. Whitehead, A. Rivers, G. Parry, J.S. Roberts and C. Button, Electron. Lett. 25:985 (1989)

51    K.K. Lau, R.H. Yan, J.L. Merz and L.A. Coldren, Appl. Phys. Lett. 56:1886 (1990)

52    J.E. Zucker, K.L. Jones, B.I. Miller and V. Koren, IEEE Photonics Technol. Lett. 2:32 (1990)

53    J.E. Zucker, M. Wegener, K.L. Jones, T.Y. Chang, N. Sauer and D.S.Chemla, Appl. Phys.Lett. 56:1951 (1990)

54    D.A.B. Miller, D.S. Chemla, P.W. Smith, A.G. Gossard and W. Wiegmann, Optics Lett. 8:477 (1983)

55    H.-C. Lee, A. Kost, M. Kawase, A. Hariz, P.D. Dapkus and E. Garmire, IEEE J. Quantum Electron. QE-24:1581 (1988)

56    E. Rosencher, Proc. SPIE Conference on Non-Linear Optical Materials III, The Hague, 1990

57    E. Rosencher, P. Bois, J. Nagle, E. Costard and S. Delaître, Appl. Phys. Lett. 55, 1597 (1989)

58    E. Rosencher, P. Bois, B. Vinter, J. Nagle and D. Kaplan, Appl. Phys. Lett. 56:1822 (1990)

59    E. Rosencher, P. Bois, J. Nagle and S. Delaître, Electron. Lett. 25:1063 (1989)

60    A. L. Lentine, H. S. Hinton, D.A.B. Miller, J. E. Henry, J. E. Cunningham and L. M. F. Chirovsky, Appl. Phys. Lett. 52:1419 (1988)

61    S. Beaumont, this volume p.

62    T. Takagahara, Phys. Rev. B36:9293 (1987)

63    E. Hanamura, Phys. Rev. B37:1273 (1988)

64    G. W. Bryant, Phys. Rev. B37:8763 (1988)

65    M. Kohl, D. Heitmann, P. Grambow and K. Ploog, Phys. Rev. Lett. 63:2124 (1989)

66    A.Nakamura, H. Yamada, T. Tokizaki, Phys. Rev. B40, 8085 (1990)

67    T. Itoh, M. Furumiya, T. Ikehara and C. Gourdon, Solid State Commun. 73:271 (1990)

68    A. Nakamura, T. Tokizaki, T. Kataoka, N. Sugimoto and T. Manabe, CLEO 90, Tech. Dig. Ser. 7 (0SA, Washington D.C., 1990) p. 178

69    A. I. Ekimov, I.A. Kudryavtsev, M.G. Ivanov and A. L. Efros, J. Lumin. 46, 83 (1990)

70    M. Kohl, D. Heitmann, P. Grambow and K. Ploog, in "Condensed Systems of Low Dimensionality", NATO ASI Series, Plenum, London (1991)

71    F. de Rougemont, R. Frey, P. Roussignol, D. Ricard and C. Flytzannis, Appl. Phys. Lett. 50:1619 (1987)

72    Y. Arakawa and H. Sakaki, Appl. Phys. Lett. 24, 195 (1982)

73    M. Asada, Y. Miyamoto and Y. Suematsu, IEEE J. Quantum Electron. QE-22:1915 (1986)

74    K. J. Vahala, IEEE J. Quantum Electron. QE-24:523 (1988)

75    Y. Arakawa, in "Waveguide Optoelectronics", NATO ASI Series Physics, Plenum (1991)

76    E. Kapon, S. Simhony, R. Bhat and D.M. Hwang, Appl. Phys. Lett. 55:2715 (1989)

77    Y. Arakawa, H. Sakaki, M. N. Nishioka, H. Okamoto and N. Miura, Jpn. J. Appl. Phys. 22:L804 (1985)

78    Y. Arakawa, K. Vahala, A. Yariv and K. Lau, Appl. Phys. Lett. 47:1142 (1985)

79    T. T. J. M. Berendschet, H. A. J. M. Reinen, H.A. Bluyssen, C.P. Harder and H.P. Maier, Appl. Phys. Lett. 54:1827 (1989)

80    D. A. B. Miller, D. S. Chemla and S. Schmitt-Rink, Phys. Rev. B-33:6976 (1987)

81    L. Banyai, Y. Z. Hu, M. Lindberg and S. W. Koch, Phys. Rev. B38:8142 (1988)

82    S. Schmitt-Rink, D. A. B. Miller and D. S. Chemla, Phys. Rev. B35:8113 (1987)

# HOT ELECTRON DEVICES

M. Heiblum[†]

IBM Thomas J Watson Research Centre
PO Box 218, Yorktown Heights
NY 10598, USA

## SUMMARY

The tunnelling hot electron transfer amplifier (THETA) structure generates an almost monoenergetic, variable energy, hot electron beam (by tunnelling), which traverses a thin GaAs region to be eventually collected and energy analyzed. As the hot electrons traverse the device they are used to probe: scattering events, band non-parabolicity, size quantization effects, intervalley transfer, quantum mechanical reflections, and band discontinuities at interfaces.

Experiments with these structures have demonstrated true ballistic transport in GaAs. Effects due to size quantization, (also expected to occur in many devices as their dimensional shrink below 100 nm or so) result as a direct consequence of the coherency maintained by the ballistic electrons. From these observations we can derive basic properties of GaAs. Speculating on the future attainable performance of the THETA device, i.e. gain and speed, we believe it has promise as a sub-picosecond amplifier.

Electronic transport in a high mobility 2D electron gas (2DEG) over distances shorter than the total mean free path (mfp) has recently attracted considerable interest. Most of the research however was concerned with 'cold electron' transport at the surface of the Fermi disc. In the present paper we report on recent experiments exploring the transport of quasi ballistic, hot electrons with excess energy up to approximately 100 meV above the Fermi level of the cold electrons background.

The study was done employing a novel three terminal device analogous to the hot electron device used in recent years to investigate hot electron transport perpendicular to the epitaxial layers, and to establish electron ballistic motion in that domain.

A generic structure is depicted in Fig.1. Two metallic gates (light areas in Fig.1) were defined employing electron beam lithography on the surface of a GaAs - $Al_{0.3}Ga_{0.7}As$ heterostructure containing a 2DEG (in the heterojunction between

---

† *present address : The Weitzmann Institute , Rehovetot, Israel*

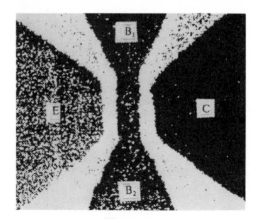

Figure 1

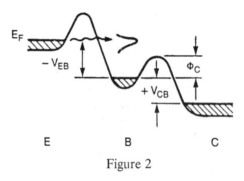

Figure 2

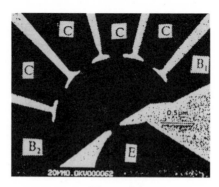

Figure 3

GaAs and AlGaAs). One gate was utilized to produce a barrier for use as a hot electron emitter (E) and the second one was used to produce a spectrometer (or collector) barrier for analyzing the energy distribution of the collected beam (C). The metallic gates were usually 500 Å wide (from left to right), the emitter gate length (top to bottom) varied between 0.25 - 1.0 $\mu$m, and the collector gate length varied between 0.75 $\mu$m and 1 $\mu$m. The separation between the emitter and the collector, the base region, ranged in the various devices between 50 and 170 nm. The various 2DEG regions, namely, E, $B_1$, $B_2$ and C were contacted by standard NiGeAu alloyed ohmic contacts. The carrier density and mobility of the different 2DEG used in the various experiments were measured at 4.2K using the standard van der Pauw procedure and were found to be 2 - 3 x $10^{11}$ cm$^{-2}$ and 3 - 8 x $10^5$ cm$^2$/V-sec, respectively, leading to a Fermi energy of $E_F \cong$ 7 - 11 meV and a transport mean free path (for cold electrons) of $l \cong$ 2.5 - 4.5 $\mu$m.

An application of a sufficiently negative gate voltage with respect to the 2DEG in the base forms an electrostatic barrier under each gate separating thus the emitter and collector regions from the base (Fig.2). Hot electrons were injected by applying a large enough voltage across the emitter barrier ($V_{CB}$ in Fig.2). Monitoring the resistance between two base contacts $B_1$ and $B_2$, on both sides of the base we ensured that the base was not depleted of carriers for the gate voltages used in the experiment. Energy spectroscopy of the current carrying states was performed via monitoring, for a given injection energy, the collected current as a function of the collector barrier height, similar to the spectroscopy mode in vertical devices. Since tunnelling through the collector barrier is negligible, only hot electrons with longitudinal (i.e. perpendicular to the collector) energy larger than the collector barrier height are collected. Electrons which relax to energies below the collector barrier height are not collected and are drained through the base to ground. This technique constitutes an almost unique way for discriminating non-ballistic transport against a ballistic one. A somewhat more elaborated version of this device (Fig.3) has been utilized to investigate the spatial distribution of the injected electrons .

A series of experiments establishing directly, via energy spectroscopy of the transmitted electrons, ballistic transport over distances as long as 0.5 $\mu$m has also been undertaken. Tunnelling through an electrostatic potential induced by a metallic gate deposited on the surface of the modulation doped structure is demonstrated and geometrical considerations are discussed. The energy dependence of the hot electron transport has been investigated. It has been shown that the main scattering mechanism at 4.2K, for hot electrons with excess energy larger than the longitudinal optical (LO) phonon energy ($\hbar w_{LO}$ = 35 meV) is LO phonon emission. For electrons with energy below 36 meV we demonstrate a surprisingly, long inelastic mfp of the order of 2 $\mu$m. This results in a previously unobserved, periodic oscillation in the maximal energy of the collected electrons as a function of the energy. In addition we have studied the angular distribution of electrons injected from a point source in 2DEG and demonstrated the steering of hot electrons by controlling the electrostatic profile of the emitter.

# HOT ELECTRONS AND DEGRADATION EFFECTS IN FET DEVICES

F. Koch

Technische Universität München
Physik-Department E16, Garching
Germany

We consider various types of field-effect devices under operating conditions to show that carriers are heated to substantial energies above kT. It is shown that light emission from the hot electrons provides a measurement of their energy distribution. We discuss degradation effects and show how the spatial location of inferface defects may determined.

## 1. INTRODUCTION

The field-effect transistor (FET) is the workhorse of modern microelectronics. By the tens of millions these rugged and reliable electronic switches are built into the advanced technology memory circuits. They come in several different versions and are made of the marvelous new compound electronic materials called $Si-SiO_2$, metal-GaAs and GaAs-AlGaAs. In this order they are familiar as the MOSFET, MESFET and the HEMT.

A common and basic element of the family of FETs is the existence of an interfacial surface between the two dissimilar materials. For the Si-based transistor it is the bounding surface between the crystalline Si and the amorphous $SiO_2$. For the MESFET it is the Schottky barrier metal-semiconductor interface. The high electron mobility transistor (HEMT) has as its bounding surface the GaAs-

AlGaAs heterointerface. The boundary asymmetrically confines the mobile charge to one side. The field-effect principle of electronic action is the control of the density of mobile charges and thus the lateral conductivity near the interface. In Fig.1 we show a cross-sectional view of a Si MOSFET.

In order to increase the electronic performance of integrated circuits, FETs have been packed ever more densely by making them smaller. The channel length of state-of-the-art Si devices is only 0.3 $\mu$m. In order to quickly switch currents on and off high lateral Source(S)-Drain (D) electrical fields are employed. The interfacial currents that flow must be of the order of 1 ma in order to quickly charge the memory capacitor with its typical pf ($10^{-12}$ Farad) capacitance. The required drain-voltage $V_D$ is in the 1-10 V range and channel widths of order 1 $\mu$m are common.

In the following we want to show that as a result of these electrical requirements the channel electrons are accelerated by huge electrical fields and become substantially hotter than ambient temperature. The carrier transport is under very non-ohmic conditions. The local density of electrical power dissipation reaches "astronomical" values. Because of the energetic carriers the boundary surface is slowly modified and device-performance degrades. We come to realize that the interfacial surface is the Achilles heel of the FET-type of device. We will discuss ways and means of how the hot carriers can be detected experimentally and their energy distribution measured. We take a look at experiments designed to measure the degradation that results from hot carriers.

## II. LOCAL ENERGY DENSITIES IN A FET. HOW HOT IS HOT?

To gain an understanding of the magnitude of the carrier heating problem we make a few simple estimates.

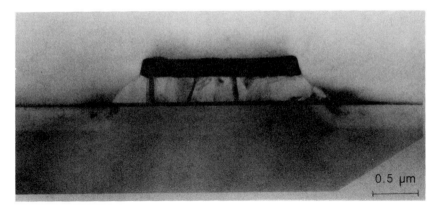

Fig.1 TEM cross section of a Si MOSFET with $TaSi_2$/poly-Si gate and selectively etched source/drain regions.

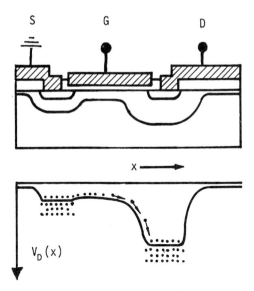

Fig.2 Schematic drawing of the potential variation $V_D(X)$ along the channel. The carrier density and conductivity decrease with x. The electric field peaks near D.

First, the lateral electrical fields that accelerate the carriers. Applying a voltage $V_D$ = 6 V over a channel length of ≈ 0.3 $\mu$m gives the average field as $E_{SD}$ = 0.2 x $10^6$ V/cm. A basic fact of FET-life is that this field is not at all constant. Because of the voltage drop along the channel, the local relative voltage $V_G$ - $V_D(x)$ decreases. It will even change sign when $V_D$ exceeds $V_G$. As a result the channel conductivity decreases with distance x away from the source. The lateral fields must increase accordingly to keep the current constant. In devices the field peaks near the drain electrode and easily reaches $10^6$ V/cm values. Schematically this is shown in Fig.2.

In mega Volt/cm fields the average carrier velocity saturates at a value near $10^7$ cm/sec. The carriers dissipate as much energy as they acquire by their motion in the field. Transport is strongly non-ohmic. The energy per carrier is of order 0.1 eV, which is substantially more than kT.

A second interesting fact relating to the energy dissipation is that the channel charges are confined by the gate voltage to a depth of order 100 Å from the interface. This depth is easily calculated from arguements related to quantum confinement of a ground state subband. Given this depth, the current density in a 1 $\mu$m wide channel is estimated as 1 ma/($10^{-4}$ x $10^{-6}$)$cm^2$ ≈ $10^7$ amps/$cm^2$. The spatial density of dissipation j · E is of the order of Terawatt/$cm^3$. In just one $cm^3$ of active semiconductor volume an amount of power is being dissipated that is of the same order of magnitude as the worldwide production of electrical energy. The energy is deposited in the drain contact region when carriers are scattered and slowed down.

Considering the operating transistor as a small point source (area ≈ 3 x $10^{-9}$ $cm^2$) from which heat is flowing into the semiconductor substrate we calculate the FET temperature as a modest ≈ 100 K above ambient. The high thermal conductivity of crystalline Si keeps the heating problem manageable. When considering the dynamic on-off operation of a memory circuit, currents are switched in a nsec and the refresh cycle is typically msec. This gives a duty cycle of only 1 in $10^6$. There are thermal transients to be accounted for. Steady state operating temperatures are reached in a time of order 0.1 $\mu$sec. The intradevice equilibration time is as short as a nsec and thus comparable with the charging time itself.

## III. MEASURING THE HOT ELECTRON ENERGY DISTRIBUTION

An experimental means of learning about hot carriers in a device is light emission spectroscopy. The electrons and holes accelerated and heated by the MV/cm electric fields inside the device are out of thermal equilibrium with their surroundings. They will radiate with a spectrum that reflects their energy distribution. This principle had been recognized for some time. It had been successfully applied to work with large area planar p-n junctions. In particular, hot carriers and the avalanche generation of minority carriers in reverse-biased junctions had been studied via the spectral distribution of the emitted light. The light emitting diode (LED) which is operated in a forward bias mode and thrives on the recombination of electron and holes is a well known device.

The dominant radiative emission mechanism is electron-hole recombination. For the unperturbed, defect-free semiconductor it is the only efficient mode of light generation. Under forward bias in a p-n junction there is massive injection of minority carriers which recombine to generate light. Such light has a spectrum which starts at the band-gap energy $\epsilon_g$. If the carriers are hot there is a tail in the energy distribution on the high side. This is a typical signature of hot-carrier recombination luminescence. The predominance of recombination processes in the light emission has made it easy to study avalanche generation of minorities and bipolar transport under forward bias.

A MOS-device operated in an appropriate range of voltages will also emit light. We show in Fig.3 a photographic exposure of a wide channel FET. A stripe of radiation is seen to emerge on the drain side where the high electric fields exist. In trying to use the spectral information available in this light from a basically unipolar device the question as to the microscopic origin must be addressed. If the light were caused by minority carrier processes it would be a diagnostic tool of little value. It would image the energy distribution of a few avalanche generated minorities in the channel. Because these minorities move out of the channel region and recombine also in other regions of the device it is not simple to extract the relevant hot carrier information from this light. It was pointed out in a little known publication of Figielski and Torun /1/, that there is expected also a sizable radiative component from Bremsstrahlung. This is the radiation expected from the majority channel electrons as they are scattered and lose their energy at the drain contact region. Because the predominant heating of carriers occurs at the position of the electric field spike Bremsstrahlung has the desired spectral information content.

In searching for radiation from Si-MOSFETs Toriumi and his coworkers /2/ used a sensitive photomultiplier. They obtained spectrally resolved data from 1.5 to 2.5 eV, far above the $\epsilon_g$ of 1.12 eV. It is not at all clear whether this is Bremsstrahlung or hot carrier recombination luminescence. The small size of the active emission area in a MOS-device makes it a difficult experiment.

Recognizing the importance of sub-bandgap measurements our work has concentrated effort on the infrared spectral region. The detector is a North Coast E 817 Ge-diode. We use a Bomem DA 3 Fourier transform spectrometer. Earlier work had used a grating instrument. In Fig.4 we show the spectrum for a Si(100) MOSFET operated with $V_D = 7.8$ V for a gate-voltage $V_G = 5.5$ V and for $T = 4.2$ K. The low temperatures allowed the identification of distinct phonon

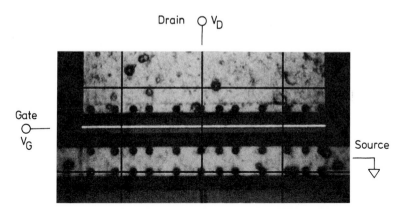

Fig.3    Microscope photograph of a wide channel Si MOSFET.
        Light emerges on the drain side of the gate
        electrode.  It is visible as a bright stripe..

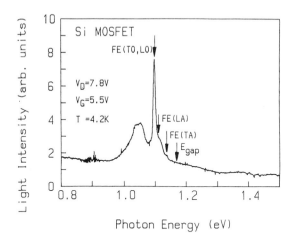

Fig.4    Emission spectrum measured at 4.2K showing several
        sharp features of recombination luminescence.

replica which are labelled here as (LA), (TA) and (LO, TO). The identification of the broad peak centered at 1,04 eV is discussed in detail in ref./3/. It is caused by holes recombining inside the drain contact. There appears to be a smoothly varying Bremsstrahlung background. Sufficient intensities of emission were achieved only for $V_G < V_D$, a condition which implies pinch-off operation of the MOSFET. In spite of the fact that we are measuring well below $\epsilon_g$ it appeared that minority carrier recombination was dominating.

Ref./4/ confirmed that suspicion. Further studies made perfectly clear that the supposedly smooth background contained broad spectral features (Fig.5). The sharp spike that emerges for $V_D$ just above 6.0 V is the recombination of cold holes and electrons in the bulk. It indicates parasitic bipolar action. It is superposed on a broad recombination peak extending to ≈ 1.2 eV and indicating hot carriers with energies of ≈ 0.1 eV and higher. At 0.98 eV appears the band-gap narrowed luminescence from holes recombing in the drain contact. The peak at 0.82 eV is most likely an intraband hole transition. In any case it is probably the same spectral feature reported by Haecker /5/ from his studies of reverse-biased p-n junctions. It must be remembered that the spectrum can be altered by self-absorption, that there is shadowing by the gate electrode and diffraction as the light emerges from the narrow slit. It is evident that Si is not a good case for Bremsstrahlung. The small $\epsilon_g$ and the lack of strong optical phonon coupling make avalanche generation highly probable. As a result minority carriers are involved in the emission of light.

It appeared to us that GaAs MESFETs are a more favorable case for Bremsstrahlung and the application of the theory of Figielski and Torun /1/. Moreover for such devices there are available Monte-Carlo simulations that give the electron energy distribution. Using the same apparatus we record in Fig.6 emission spectra that show a continuous, smooth exponential decrease. The curves are described by an

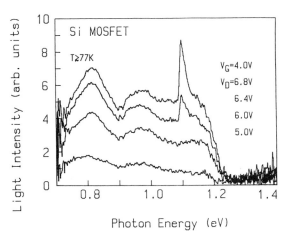

Fig.5 Spectra as a function of $V_D$ for fixed $V_G$. Distinct spectral feature are observed well below $\varepsilon_g$.

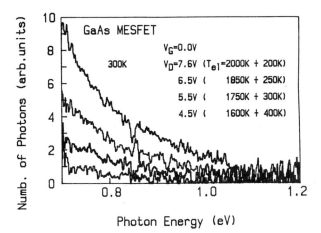

Fig.6 Emission spectra for a GaAs MESFET device. The number of photons emitted decreases smoothly and exponentially as expected for Bremsstrahlung. The decay constant $T_{el}$ give a measure of the electron energy.

decay constant which relates to the electronic energy in the FET channel. Expressed as a temperature, values of order 2000 K are found from the spectra.

In ref./5/ is discussed how the measured emissions compare with the device simulation and the theory of Bremsstrahlung. Suffice it to say here that the comparison is convincing. With regard to GaAs MESFETs we can now state that spectroscopy of the light emission gives an experimental measure of the energy distribution of hot carriers.


## IV. DEGRADATION EFFECTS

The $Si-SiO_2$ interface is a network of some $10^{15}$ chemical bonds per $cm^2$ connecting O and Si atoms. Because of the randomness of the structure there will be a spectrum of good and bad bonds, some severely strained and distorted, others near perfect. Statistically speaking there will be situations where things just don't match up. It is well known that after thermal oxidation there remain typically $10^{12}$ electronically active trap states per $cm^2$. Annealing the structure in the presence of hydrogen is able to reduce this density again by one or two orders of magnitude. Current thinking attributes to hydrogen the ability to passivate bonds and thus cure the electronic ills. In Fig.7 is shown a high resolution TEM picture of a typical interface.

Energetic electrons in high electric fields will in time be able to cause damage, perhaps even remove the hydrogen used to patch up a poor bond. Hot electrons can be injected and trapped inside the amorphous $SiO_2$. In general the interface will degrade with the time of exposure to the hot electrons. A typical signature of this degradation is the drift of the transistor threshold voltage to higher positive values. It appears that negative charge is trapped and a barrier potential built up that prevents the transport of channel electrons until a higher gate voltage

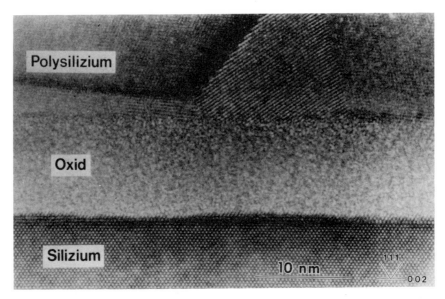

Fig.7   HRTEM photograph of a typical Si-SiO$_2$ interface.

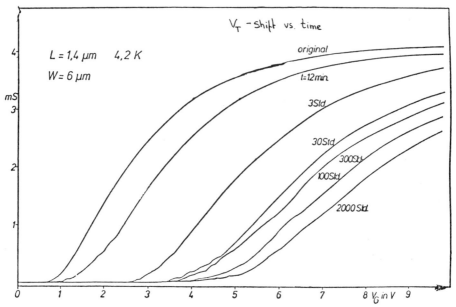

Fig.8  Transconductance characteristic measured at 4.2K after
various times of stressing with a high source-drain
electric field.

is reached. We show in Fig.8 an example of how electrical stressing alters the transconductance curve $I_D$ vs. $V_G$.

Because of the field spike it is expected that the damage is strongly peaked near the drain. Here where the field is highest, where the light emerges and electrons are heated most strongly is expected that the damage is concentrated. Simulation modelling is consistent with this assumption. Device models which locate the extra interface defects within a distance of $\approx 0.1$ $\mu$m from the drain can satisfactorily account for the electrical characteristics. The defects, it is believed, lie in a narrow strip across the entire width of the transistor.

There are quite a few ways in which the defect states have been studied. One has observed tunneling conductance through the potential barrier formed by the charged defects /6/. Switching noise from statistical occupation fluctuations in that strip have been explored /6,7/. The temperature dependence of subthreshold current has been modeled /8/. None of these, however, could give precise information on the location of the defects along the interface.

In this final part of the paper I want to describe a recent experiment that has provided the required information. The technique elegantly combines device modelling and measurements in a way that is exemplary for the analysis of the small microstructures. It also shows that defects caused by hot electrons act as generation centers.

The experiment /9/ uses of the fact that when the drain is biased positively a small reverse diode current will flow through the substrate to ground. The current in a Si-diode is primarily caused by generation of carriers at defects inside the depletion layer. Shockley-Read-Hall statistics assure us that the generation is exponentially weighted in favor of those defects that are located at the mid-gap energy position. Provided that $eV_D$ is much smaller than $\epsilon_g$, the effective generating centers lie in the middle of the depletion layer that surrounds the drain.

Fig.9 shows the generation zone as a region of space between the two energy contours $\psi_e^i$ and $\psi_h^i$. The figure makes clear that the shape of the depletion region, and with it the generation zone, can be adjusted with the application of the gate voltage. In the jargon of microelectronics, this is a gated-diode configuration. Careful modelling of the potentials, the doping profile and physical dimensions of the structure are required to know exactly where the generation zone contacts the $Si\text{-}SiO_2$ interface. As the gate voltage is swept from negative to positive values, the zone moves like a pointer along the interface.

The experiment procedes to measure the generation current $I_{gen}$ as a function of $V_g$. For the undamaged FET the curve of $I_{gen}$ vs. $V_g$ is rather flat and featureless for negative bias. Near $V_g = 0$ the current rises to form a peak as the generation zone sweeps along the interface. For positive $V_g$, the current again settles at a constant value.

The sample is then subjected to hot electron stressing by operating the FET for some length of time at high $V_D$. After doing so, the $I_{gen}$ vs. $V_G$ characteristic is again recorded. As in Fig.10 there is a clear and easily measured increase in $I_{gen}$. In particular, the peaking at negative $V_g$ shows that the extra defects are located at the $Si\text{-}SiO_2$ interface and near the drain electrode. The insert shows a simulation modelling of $\Delta I_{gen}$ in which the peak position of the distribution of defects is moved about in steps of 10 nm. The best fit is found for the peak position of 4 nm to the left of the metallurgical drain-substrate p-n junction A width of 0.1 $\mu m$ for the distribution of defects is required to fit the curve.

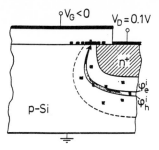

Fig.9    Depletion  layer  and  generation  zone  for  reverse-bias
        gated-diode operation of the MOS transistor.

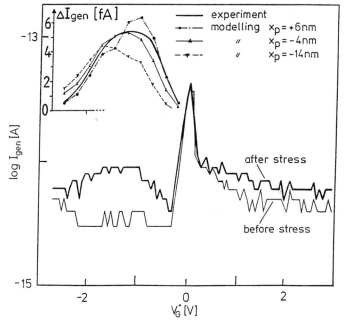

Fig.10   Diode generation current as a function of gate voltage.
        The  increased  $I_{gen}$  after  stress  is  caused  by  interface
        defects located near the edge of the drain contact.

## V. CONCLUDING REMARKS

   Hot  carriers  are  a  fact  of  life  for  field-effect  devices
under typical operating conditions. Their analysis by means
of   emission   spectroscopy   appears   possible   when   Bremsstrahlung of the majority carriers can be observed.

   The  interfacial  surface  of  a  Si  MOSFET  sensitively
reacts  to  hot  carrier  damage.  Similar  effects  must  be
expected   for   the   Schottky-barrier   of   a   GaAs   MESFET.

Possibly the epitaxially perfect heterointerface of the GaAs-AlGaAs is less susceptible to damage. We recognize that the interfaces in a FET-device are most easily affected by hot carriers. Here is where degradation occurs. The defects appear concentrated where the carriers have been heated most strongly.

**REFERENCES**

/1/ T. Figielski and A. Torun, On the Origin of Light Emitted from Reverse Biased p-n Junctions, in: Proc. of the Int. Conf. on the Physics of Semiconductor, Exeter, 863, 1962

/2/ A. Toriumi, M. Yoshimi, M. Iwasa, Y. Akiyama and K. Taniguchi, A Study of Photon Emission from n-Channel MOSFETs, IEEE Trans. on Electron Devices, ED-34:1501 (1987)

/3/ M. Herzog and F. Koch, Hot-Carrier Light Emission from Silicon Metal-Oxide-Semiconductor Devices, Appl. Phys. Letters, 53:2620 (1988)

/4/ M. Herzog, M. Schels, F. Koch, C. Moglestue and J. Rosenzweig, Electromagnetic Radiation from Hot Carriers in FET Devices, Sol. State Electronics, 32:1065 (1989)

/5/ M. Herzog, F. Koch, C. Moglestue and J. Rosenzweig, Proc. GaAs and Related Compounds 1990, to be published

/6/ M. Bollu and F. Koch, Quantum Transport Effects in Small Si-MOS Devices, Proc. 18th Int. Conf. on the Physics of Semiconductors (Stockholm), p. 1519, World Scientific (1987)

/7/ M. Bollu, F. Koch, A. Madenach and J. Scholz, Electrical Switching and Noise Spectrum of Si-SiO$_2$ Interface Defects Generated by Hot Electrons, Appl. Surf. Sci., 30:142 (1987)

/8/ A. Asenov, M. Bollu, F. Koch and J. Scholz, On the Nature and Energy Distribution of Defect States Caused by Hot Electrons in Si, Appl. Surf. Sci., 30:319 (1987)

/9/ P. Speckbacher, A. Asenov, M. Bollu, F. Koch and W. Weber, Hot-Carrier-Induced Deep-Level Defects from Gated-Diode Measurements of MOSFETs, IEEE Electron Device Letters. 11:95 (1990=

# LECTURERS

G. Abstreiter
Walter Schottky Institut
Technische Universität München
Am Coulombwall
D-8046 Garching
FRG

D. Anderson
Royal Signals & Radar Establishment
St Andrews Road
Great Malvern
Worcs WR14 3PS
UK

G. Bastard
Ecole Normale Supérieure
Département de Physique
24 rue Lhomond
75231 Paris cedex 05
France

S.P. Beaumont
Nanoelectric Research Centre
Deptment of Electrical Engineering
University of Glasgow
Glasgow G12 8QQ
UK

W. Beinvogl
Siemens AG
Otto-Hahn-Ring 6
D-8000 München
FRG

H.G. Grimmeiss
Dept. Solid State Physics
University of Lund
Box 118
S-221 00 Lund
Sweden

B. Hamilton
Dept. Pure & Applied Physics
UMIST
PO Box 88
Manchester, M60 1QD
UK

M. Heiblum
IBM
Thomas J Watson Research Center
PO Box 218
Yorktown Heights NY 10598
USA

M. Jaros
School of Physics
Dept. Theoretical Physics
University of Newcastle
Newcastle upon Tyne NE1 7RU
UK

F. Koch
Physik-Department E16
Technische Universität München
D-8046 Garching
FRG

M. Leys
Technische Universiteit Eindhoven
Den Dolech 2,
PO Box 513
5600 MB Eindhoven
The Netherlands

J.L. Merz
Center for Electronic Structures
University of California
Santa Barbara, CA 93106
USA

A. Ourmazd
AT&T Bell Laboratories
Crawfords Corner Road
Holmdel
NJ 07733-1988
USA

A.R. Peaker
Centre for Electronic Materials
UMIST
PO Box 88
Manchester M60 1QD
UK

K. Ploog
Max-Planck-Institut für
Festkörperforschung
Heisenbergstrasse 1
Postfach 80 06 65
87000 Stuttgart 80
FRG

C. Weisbuch
Thompson-CSF
Laboratoire Central
Domaine de Corbeville
91404 Orsay
France

# INDEX